Black Team

Formation 'arrow' low-level flight of no less than sixteen Hunters from the 111 Sq. aerobatic team.

HAWKER HUNTER

The story of a thoroughbred

Introduction

The most successful British jet fighter produced was without doubt the sleek and graceful Hawker Hunter. As usual for every new aircraft type it had its share of teething problems, but when these were all adequately solved the U.K. had at that time one of the best jet fighters available. It was built in large numbers and exported to many countries. As a military fighter the Hunter is now retired for many years but even today several airworthy Hunters are still present in the air-show circuit! With their easy maintenance and straightforward construction they'll hopefully continue to be the star at various air-shows in the coming years!

P.1052

The P.1052 was the first step towards the Hunter: swept wings but still a straight tail.

A step-by-step development from the Hawker Seahawk

Shortly after the war, Hawker produced for the Royal Navy the Hawker Seahawk jet fighter fitted with straight wings and a Rolls Royce Nene centrifugal jet engine. As first step towards the Hunter, the Seahawk was fitted with 35° swept wings as the Hawker P.1052. However it still had the same tail as the Seahawk. It flew for the first time in November 1948. Later, the second P.1052 was fitted with a new rear fuselage with a swept tail as the P.1083. It still had a Rolls Royce Nene which resulted in a relatively broad fuselage. Redesigned with the new axial flow Rolls Royce Avon jet engine and with a much sleeker fuselage, this would finally

result in the P.1067, the Hunter prototype. Both P.1052 and P.1083 were unable to exceed the speed of sound in a dive. The Hunter could do this without problem! So, the Hunter was not a radical new design, but the result of a step-by-step further development of the first Hawker jet fighter, the Seahawk.

Early test flying and development

For replacement of the Gloster Meteor both Vickers Armstrong (Supermarine) and Hawker developed a fast-climbing jet fighter with trans-sonic capability that resulted finally in the Swift and the Hunter. The Royal Air Force clearly betted on two horses at that time and both types finally entered production.

The Hawker P.1067 design, following Specification F.3/48 (later replaced by Spec. F.43/46) was submitted to the Air Ministry to meet this specification. This was awarded on 14 March 1951 by a contract for the construction of three prototypes. These three aircraft received the R.A.F. serial numbers WB188, WB195 and WB202. As already discussed the P.1067 had its pedigree in two steps from the Seahawk. Also the competing Vickers Armstrong design, later known as the Swift, had a similar history when their straight wing Nene-powered Attacker naval jet fighter was fitted with swept wings and finally developed into a design with a Rolls Royce Avon engine; the same as for the P.1067.

Initially the Hawker design team, headed by

P.1081

Last step before the Hunter: the P.1081 with both wings and tailplanes swept.

WB188

WB188 was the first P.1067 prototype for the Hunter. It was unarmed and without an air–brake.

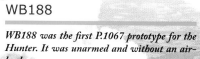
Hunter prototype

First prototype was painted in overall Sky colour, serial WB188

Type 535

The Vickers Armstrong Type 535 VV119 was the prototype for the final Swift fighter. Only 197 were built and they were soon phased out after structural airframe failures were encountered. The Swift was the alternative in case the Hunter would fail but it was the Swift that was less lucky!! They were replaced by......Hunters!

Vickers Armstrongs Supermarine 535

Sidney Camm, proposed for the new jet fighter a nose air intake, a 40° swept wing and a high tail with an armament of four 20 mm cannons, After wind tunnel tests of various configurations a low tail was selected in combination with a wing with 35° sweep and wing root air intakes of the same shape as the preceding P.1083. As armament the new 30mm Aden cannon was selected. Meanwhile, the construction of the first prototype WB188 was already started by end 1949. It made its first flight on July 20th 1951 with Hawker's chief test pilot Neville Duke at the controls. At this stage it did not yet carry any armament. WB195 was armed with four Aden 30 mm guns. It made its first flight on May 5th 1952. Both WB188 and WB195 were fitted with the new Rolls-Royce

Avon jet engine. The third P.1087 prototype was fitted with an Armstrong Siddeley Sapphire turbojet as a 'fail safe' in case the Avon engine would run into great problems. It made its first flight on November 30th 1952 carrying full gun armament.

The flight tests with WB188 showed that the new fighter had such high flight performances that it was later prepared to establish a new world speed record. For this purpose a higher rated Avon with afterburner was fitted. Also the nose was better streamlined ending in a sharp point. Sprayed in an all-over bright red colour it set an absolute world speed record of 1171 km/h on September 7th 1953, flown by Neville Duke. On September 19th of the same year, Duke also set a new world speed record

WB195

WB195, the second P.1067 prototype displayed at the Farnborough airshow in 1952. It was shown together with the Swift!

WB188

WB188 with pointed nose, rear side-mounted double airbrakes and an up-rated Avon engine with an afterburner as used for breaking the world speed record. In this form it is referred to as the Hunter Mk.3.

Sidney Camm

Sir Sydney Camm, CBE, FRAeS (5 August 1893 – 12 March 1966) was a British aeronautical engineer who contributed to many Hawker aircraft designs, from the biplanes of the 1920s to jet fighters. One particularly notable aircraft he designed was the Hawker Hurricane fighter.

Camm grew up in a large family at Windsor. He started his first design activities at the Windsor Model Aeroplane Club and followed a carpenter's apprentice course. Later he joined the Martinside aircraft company as a draughtsman. In 1923 he joined Hawker Aircraft Company at Kingston-Upon-Thames where his qualities were soon recognized. Already two years later he became Chief Engineer. He was active as an engineer at Hawker, later Hawker Siddeley, until his retirement. Camm was also active at the Royal Aeronautical Society. He died in 1966 at the age of 72.

WB202

A beautiful in flight photo of the third Sapphire powered Hunter prototype WB202.

State Express

*WT155 was the very first Hunter Mk.1 pro-
duction model, still without an airbrake.*

Tropical trials

*The Mk.1 Hunter WP569 was used in 1954 for
trials under tropical conditions. On this photo
it is shown at Khartoum airport, Sudan with a
makeshift sun shade for the comfort of the pilot.*

over a 100 km closed circuit at 1141 km/h.

In April 1952, the Hunter prototype exceeded
the speed of sound in a shallow dive with Duke
at the controls. Except for some vibration at the
trans-sonic speed range the Hunter flew very
well. It had a high manoeuvrability and could
be safely spun as Duke showed several times at
air displays. Initially the Hunter flew without an
airbrake, but later on this device was fitted on all
Hunters under the rear fuselage. After its evalua-
tions tests were completed WB188 was used for
the world record speed flights with an up-rated
Avon engine and two side mounted airbrakes.
WB188 ended its days as an instructional airframe at Halton.

WB195 was used for airbrake development and
initial spinning trials. Later, it became an instructional airframe as well, registered 7284M.
WB202 was also used for airbrake development
and for drop tank trials at Dunsfold, Hawker's
new aerodrome. It was later fitted with four dum-
my Firestreak air-to-air missiles. It was scrapped
in 1960.

Measurements

*Hunter F.1 WT571 was used for drag and lift
measurements at speeds up to Mach 1.2. For
this purpose it was fitted with a long nose probe
for pressure measurements in undisturbed
air. All recording equipment was fitted in the
weapons bay in the nose.*

T.7 of No. 74 Sq RAF

Unit insignia of this all silver machine was tiger head with bars on the nose

T. 7A RAE 1982

Royal Aircraft Establishment machine XF321 in Raspberry Ripple pattern

FGA.9 of No. 20 Sq RAF

Standard RAF camouflaged machine, unit logo on nose

FGA. 9 of No. 43 Sq RAF

This aircraft shows worn out state of its camouflage colours with some surface completely peeled of paint. Unit insignia on nose as well rear fuselage

GA. 11 of No. 764 Naval Air Squadron

This machine served in 1968 with top surface in Extra Dark Sea Grey and bottom in white. Dorsal Spine in white as well as letters on tail and rear fuselage

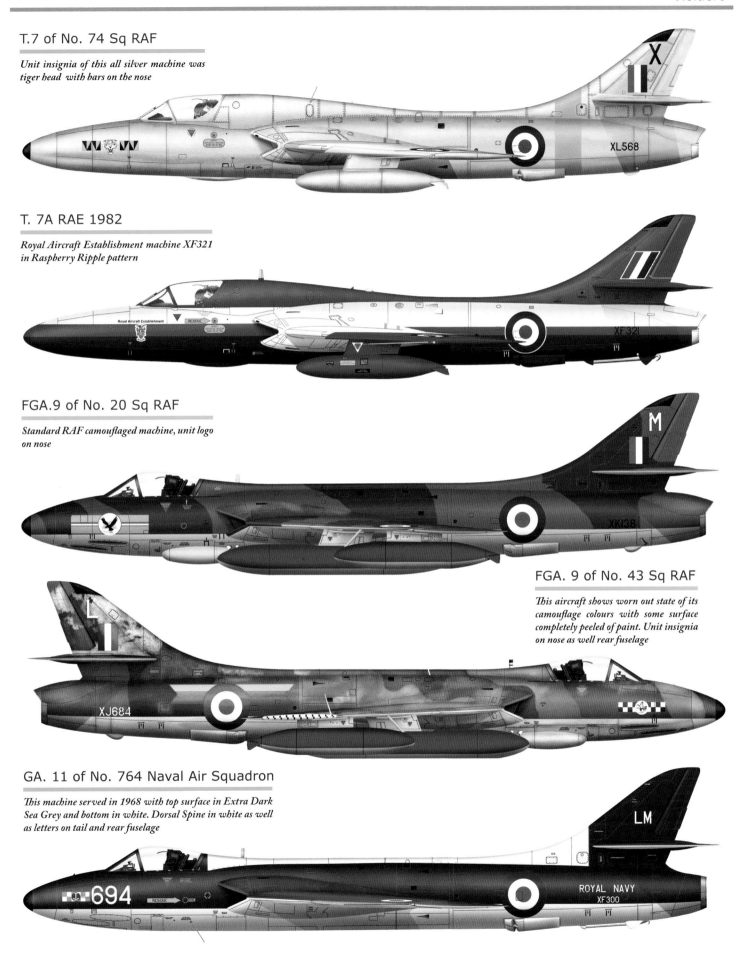

Leuchars

Hunters F.1 nos. WT582, WT622, WT618 and WW645 from No. 43 Squadron at Leuchars (with airbrakes) in a very tight and tidy formation.

Delta Jets

Hunter F.4 XE665 was operational at No. 130 Squadron based at Jever, Germany. It was later rebuilt as a T.8 two-seat trainer and still exists, owned by Delta Jets Ltd.

Fire Flash

Hunter F.4 XF310 was used for experiments with the new Fireflash air-to-air missile. The radar installation was mounted in a lengthened and well streamlined nose. Two Hunters were also converted for experiments with the Firestreak air-to-air missile as P.1109A and P.1109B.

T. 8 of FRADU

This machine operated in overall Light Aircraft Grey scheme with day-glow red surfaces under nose, dorsal spine, vertical tail and wing roots.

Into production

With only three Hunter prototypes available for test flying, the R.A.F. urgently needed production fighters that could not only replace the ageing Meteors, but that also would be on even terms with the new MIG-15 fighter. This resulted in a production order for 113 Mk.1 Hunters and 45 Sapphire powered Mk.2's. Hawker lacked the facility to built all and the Mk.2's were finally built by Armstrong Whitworth. The new Hunters started to roll from the production lines between May and October 1953.

The first Mk.1's were used for operational testing and as usual for a combination of a new airframe and a new engine teething problems soon began! During flight endurance trails it was discovered that the range of the Hunter on its internal fuel supply was hardly sufficient for a one hour combat mission excluding start and landing and gaining altitude. This was already

experienced with the prototypes that sometimes had to abort their test flights for lack of fuel. To solve this problem, two underwing drop tanks were developed with a capacity of 454 litres (100 imp. Gallon) each.

A more serious problem was the new Rolls Royce Avon engine. It was found to be susceptible to flame-out caused by surging of the first rows of compressor blades.

It took a while before Rolls Royce solved this problem by giving the first row of compressor blades a variable pitch. However, this was not the end of the surging problem. When the Hunter started its first gun firing trials the shock waves resulted again in flame-outs because of engine surging. These shock waves entering the air intake disrupted the airflow to the engine. Final solution was mounting two external bulges under the fuselage nose that dispersed the shock waves from the firing cannon. The Sapphire powered Mk.2 never had this engine surging problem and did not need

the extra bulges! Even with the two bulges cannon shock waves sometimes found their way into the air intake and as final solution small deflection plates were fitted on the end of the gun barrels.

The production of the Hunter was started in such a hurry that the first batches were not even fitted with an airbrake since this was still under testing at the prototypes! When the airbrake testing was completed, they were retrofitted as yet.

The Mk.1 and Mk.2 were soon followed by the Mk.4; the Mk.3 was the original prototype WB188 fitted with up-rated Avon with an afterburner for the world speed record flights that did not go into further production.

Hunter F Mk.1 and 4

The Mk.1 and 2 versions of the Hunter were strictly preliminary versions of the final Hunter since they were at that point, without external fuel tanks, only suitable as short range interceptor at the operational squadrons. Modi-

Lift Off

Hunter F.5 WP188 from No.56 Squadron at lift off.

Payload tester

Hunter F.5 WN958 was used by the A.&A.E.E. for investigations with various underwing stores that included unguided rockets as we can see. Strangely it was fitted with the cannon bulges of the F.4 and F.6 while these were unnecessary at this type! It was later 'demobilized' as instructional airframe No. 7397M. It was scrapped in 1960

fications to fit extra internal wing tanks and external drop tanks (and adding of the extra bulges under the fuselage to solve the gun firing problems!) were introduced at WT107, the 114th production machine from Kingston-upon-Thames. At this moment, the aircraft became known as the F.4. Also a number of earlier produced Hunters were modified as F.4. Production F.4 Hunters found their way to the various operational R.A.F. squadrons. The F.4

was also the first machine attracting foreign interest and soon F.4's were not only exported, but also licence-produced in the Netherlands and Belgium. With all teething problems now fully solved the Hunter showed its real potential and in general it showed a surprisingly good service record. Especially the easy handling and maintenance were appreciated and resulted in remarkably short turn-around times after a mission. If all went well, a Hunter could

be refuelled and rearmed in seven minutes! The Hunter F.4 continued in service with the operational squadrons until 1957 when it was replaced by the much improved Mk.6. When withdrawn from the operational fighter squadrons many F.4's found their way to conversion and training units. A number was later converted into T.7 two seat trainers. Further numbers were refurbished and exported to various countries.

T. 8M of No. 899 Sq

Royal Navy Hunter T.8 XL580 fitted with a Hawker Sea Harrier radar installation. Standard Extra Dark Sea Grey/White

KLu F. 6

Netherlands was one of the European countries operating Hunters in '60. Camouflage is standard, 'rosettes' in four positions.

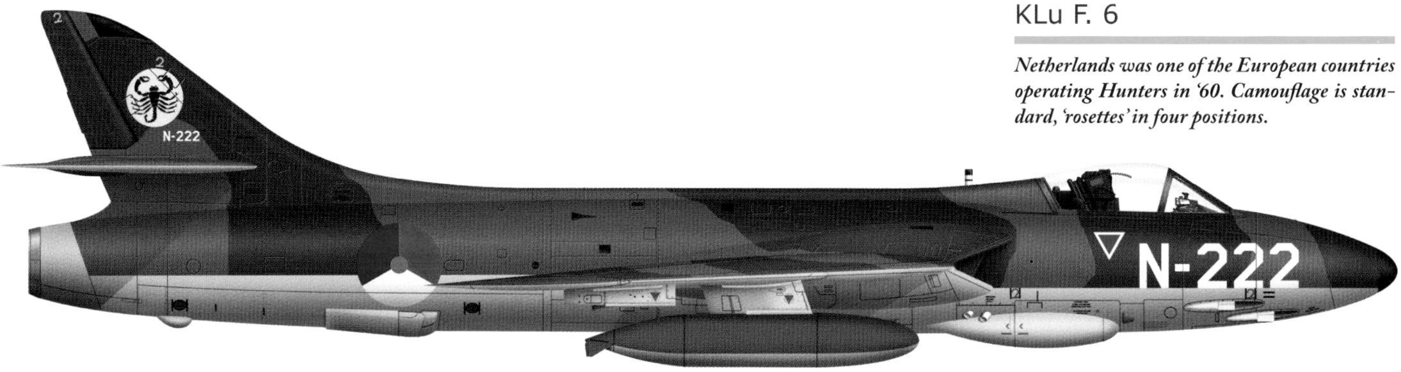

India F. 56

Machine shown here served in early '70. Note unit insignia placed at the fuselage rear as well on nose

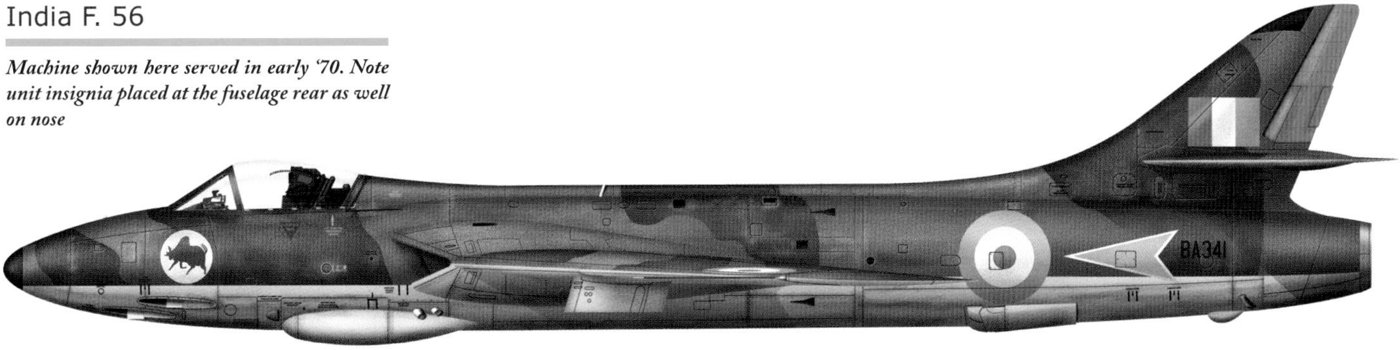

Jordan FGA. 9

In spite of serving in the Middle East, this machine wears standard European camouflage. Number on fuselage in white as well tail letter.

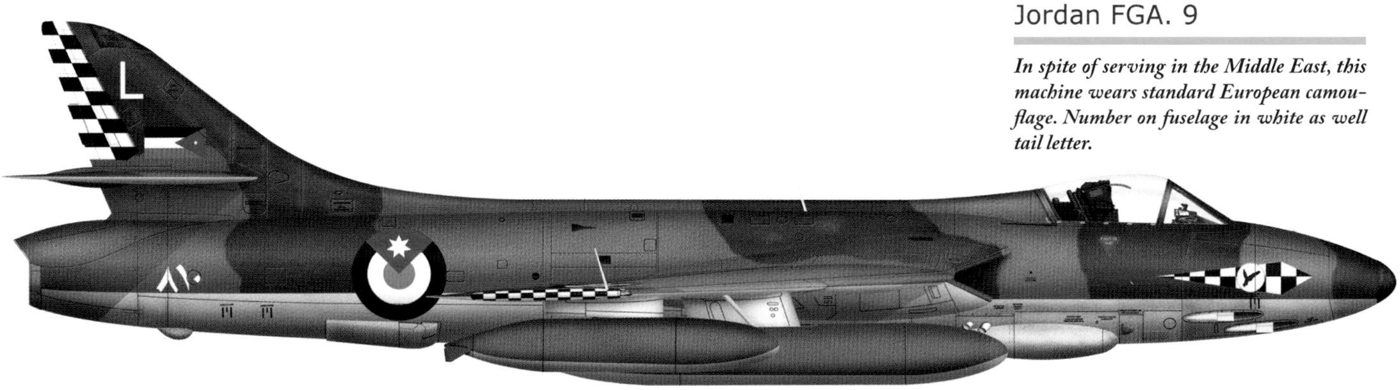

Hunter F Mk.2 and 5

The 45 Sapphire powered Mk.2 Hunters went into operational life without the problems of their Avon powered 'sisterships'. They also lacked the characteristic bulges under the fuselage nose since these were not necessary. Later a second and last batch of 105 Sapphire powered Hunters was delivered as the Mk.5 that was grossly identical to the Mk.2 but also had the extra wing fuel cells and underwing drop tanks of the Mk.4. These Hunters went into operational use without further problems until 1958,

when they were replaced by the Mk.6. The F.2's were never brought up to Mk.5 standards, but simply decommissioned and scrapped, sometimes with very low flying hours.

Hunter F Mk.6

Hawker's chief designer Sidney Camm already considered, even before the P.1067 prototype made its first flight, a supersonic version with wings with a sweep of 50°. Wind tunnel investigation gave promising results predicting that the P.1067 fitted with more swept wings, an up-rated and improved Avon engine and an

afterburner would go supersonic in horizontal flight. Eventually this supersonic P.1067 version, known as the P.1083, was cancelled by the British government at a stage when a prototype was already under construction. The new and improved Avon, known as the Avon 200 series, produced 4536 kg static thrust and with the use of a new high quality of steel for the high pressure section of the compressor it was also less susceptible to the earlier surge and flameout problems. With this new engine ready for production and the P.1083 cancelled, Camm's

XF833 was the P.1099 prototype for the much improved Hunter F.6. We see it here at Rolls Royce for thrust reversal experiments.

Team

A nice example of precision flying by the 111 Sq. aerobatic team. It takes great flying skills to perform a formation looping!

team worked on a vastly improved version of the Hunter that became finally the F.6. It received the company designation P.1099. At least the incomplete P.1083 served a very good purpose: it was used for the construction of the prototype for the new Hunter F.6. It made its first flight in an all-over silver scheme on January 22nd 1954 by Neville Duke with R.A.F. serial number XF833. It was extensively tested at Dunsfold and later at the A.& A.E.E. (Air-

craft and Armament Experimental Establishment) at Boscombe Down. When its evaluation was completed and series production of the F.6 was started, XF833 was used by Rolls Royce for thrust reversal experiments.

The final F.6 was powered by an Avon 203 giving some 30% more thrust than the earlier versions. It gave the F.6 no supersonic properties, but it could reach Mach 0.95 at altitude and 1118 km/h at sea level. Furthermore, it gave

the Hunter for its days a phenomenal rate of climb of more than 86 m/sec. Other changes with the earlier marks included an extended part of the outer wing leading edge, giving the F.6 its characteristic 'dog-tooth' appearance. The extended leading edge was introduced to give the Hunter a better turning ability at high speed turns where the earlier wing tended to give unwanted pitch changes.

The R.A.F. received a pre-production batch of

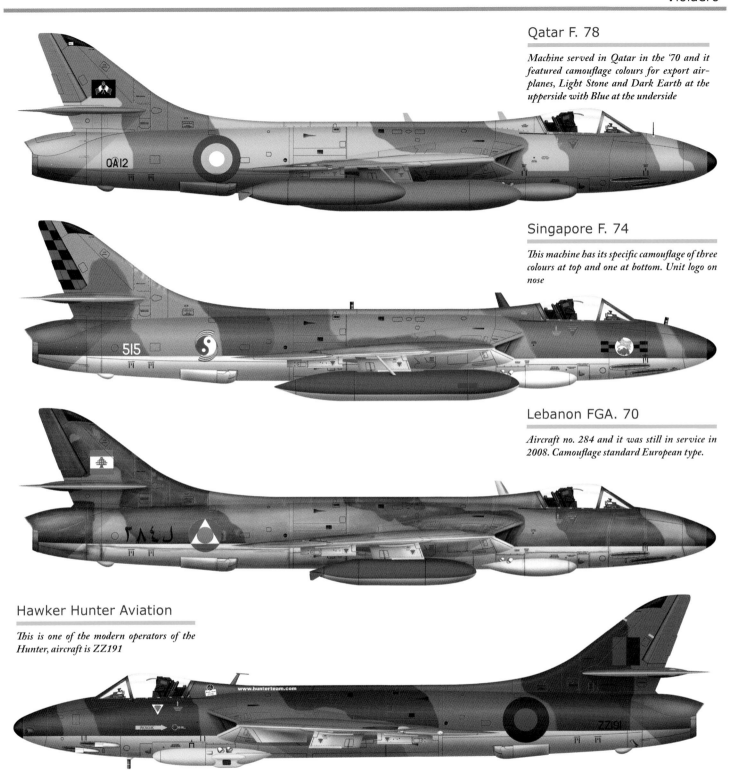

Qatar F. 78

Machine served in Qatar in the '70 and it featured camouflage colours for export airplanes, Light Stone and Dark Earth at the upperside with Blue at the underside

Singapore F. 74

This machine has its specific camouflage of three colours at top and one at bottom. Unit logo on nose

Lebanon FGA. 70

Aircraft no. 284 and it was still in service in 2008. Camouflage standard European type.

Hawker Hunter Aviation

This is one of the modern operators of the Hunter, aircraft is ZZ191

seven aircraft, followed by a further 308 produced over four batches. First F.6 delivered was WW592 that flew on March 25th 1955 for the first time. By the end of 1956 the F.6 was entering squadron service with the R.A.F. with a number also stationed in Germany. The F.6 gained fame by the many airshow displays given by No.111 Squadron with a large formation of Hunters painted in an all-over glossy black colour scheme. The display group was named the 'Black Arrows'.

The last Hunter F.6 delivered to the R.A.F. for operational use was XK156 that flew for the first time on July 9th 1957 by Duncan Simpson. Hunters F.6 XK157-XK176 (20) and XK213-XK224 (12) from the last production batch were completed but not for the R.A.F. They were later used for export orders.

At the last final order one hundred Hunter Mk.6 aircraft were cancelled as a consequence of the infamous 'White Paper' document from Duncan Sandys that foresaw the end of the manned fighter in the UK with the Lightning being the last operational type!

By December 1957 the R.A.F. had a total of 210 Hunter F.6 fighters on strength with another 159 being held in reserve at the various Maintenance Units. Phase out of the Hunter as a fighter started from 1960 on when they were replaced by the Lightning and later by Rolls Royce Spey powered F-4 Phantoms purchased from the United States. Many F.6's were later converted as ground attack fighter and served on until replaced by modern types. F.6's single seat fighters were also converted to dual seat trainers. Refurbished export F.6 Hunters were

13

Display team

A typical air display setting: five 111 Sq. Hunters in formation with a Handley Page Victor bomber.

Mixed pack

'Planes to come and planes to go'; a mutual flight of two 111 Sq. Hunters with two new Lightning fighters.

used operationally as late as the nineties by various countries all over the world. Even at present, in 2012, a number of airworthy Hunters in private hands are still giving aerobatic displays at many airshows and hopefully they will continue these displays over many years to come!

Hunter FGA. Mk.9

In the sixties the majority of R.A.F. ground attack units were equipped with the Hunter. Strangely enough none of the ground attack versions of the Hunter were newly produced aircraft. All Hunter ground attack fighters were converted from F.6 fighters. This version was designated as the FGA.9 and differed from the F.6 by a strengthened wing for offensive loads and standard two1046 1 (245 Imp Gallons) drop tanks on strengthened inner wing pylons. The FGA.9's used in tropical regions were further fitted with a brake parachute and extra cockpit ventilation.

In total 128 F.6 Hunters were converted for the R.A.F. as FGA.9 ground attack fighters and they started to join the first Strike Command squadrons (Nos. 1, 8 and 54) in 1960-1961 and were used for a period of almost nine years.

However, this was definitely not the end for the ground attack Hunter! In fact many of the Hunters exported to foreign countries were directly based on the FGA.9 and they were only adapted as wished by the customer regarding radio and navigation equipment and instrumentation.

Hunter FR Mk.10

This mark was never newly produced either, but just like the previous Mk.9 converted from existing Hunter F.6 fighters. In total 33 Mk.10's were modified. FR stood for 'Fighter

Hawker team

'Neville Duke sitting in the cockpit of the Hunter. The standing person is Hawker Siddeley test pilot Duncan Simpson'

Reconnaissance'. As fighter it retained the four Aden cannon, but in a new nose cone section three Vinten F.95 cameras were mounted, fitted with either 10 cm or 30.5 cm focal lenses. One camera was fitted in the front part of the nose for forward photography, protected by an 'eyelid' shutter when not in use. The two others were bracket mounted in a sideward position for oblique photographs. The oblique cameras were mounted in a staggered position meaning the camera windows at each side of the nose were at different heights. Further, the FR.10 was fitted with a brake parachute and the strengthened inner pylons with the larger droptanks from the FGA.9. First flight of the FR.10 prototype XF429 took place on 7 November 1958. Since this version was planned to operate at low to very low altitudes it had an extra armour plate fitted under the floor to protect the pilot against small calibre guns fired from the ground. The order for the conversion

Neville Duke

Neville Frederick Duke was born on 11 January 1922 at Tornbridge, Kent. After his education at the Convent of St. Mary and The Judd School in Tornbridge he worked as an auctioneer and estate agent. He tried to join the Fleet Air Arm on his 18th birthday, but was rejected. He successfully joined the R.A.F. as a cadet in June 1940.

After completion of his pilot training he was finally assigned to No. 92 Sq at Biggin Hill in April 1941 as a Spitfire Mk.V pilot. Until October 1941 he managed to claim two Messerschmitt Me-110's. He was posted in North Africa where he flew the P-40 Tomahawk at no. 112 Sq. On 30 November 1941 he was shot down by a German Expert, Oberstfeldwebel Otto Schulz and on 5 December of the same year he was shot down again by a JG 27 pilot. In spite of these misfortunes he managed to shoot down German planes after he was converted from the Tomahawk to the faster Kittyhawk. By February 1942 his score was already eight conformed kills. He completed his African tour by end 1942 and spent the following six months as an instructor in the U.K.

In November 1942 he joined No. 92 Sq, again in North Africa. Here he flew a tropicalized Spitfire Mk.V. He became a flight commander and ended the war with 27 confirmed kills making him one of the highest ranking U.K. fighter aces.

In January 1945 he started to work for Hawker as a test pilot, still under his military contract. He successfully attended the No.4 course at the Empire Test Pilot School at Cranfield in 1946 and finally became Hawker's chief test pilot in 1951.

Duke gained fame with his Hawker Hunter flights; not only for his world speed record flights in 1953 but also for the many airshow demonstrations he gave with the Hunter. He showed the Hunter, fitted with smoke canisters, regularly with prolonged spins. Duke ended his career as a test pilot in October 1956 after serious spinal injuries following two crash landings.

Duke took up freelance aviation consultancy work until 1960, when he formed Duke Aviation Limited. He was Sir George Dowty's personal pilot for most of the 1960s and 1970s. He sold the company in 1982. He also became a test pilot for Edgley Aircraft and later Brooklands Aircraft on the Edgley Optica and Brooklands Firemaster 65.

Duke wrote several books based on his experiences. His autobiography, Test Pilot, was published in 1953 and reprinted in 1992. His other books include The Sound Barrier (1953), The Crowded Sky (1959) and The War Diaries of Neville Duke (1995). On 7 April 2007 Duke was flying his private aircraft when he felt unwell. He landed safely at Popham Airfield, but collapsed when he left the aircraft. He was taken by ambulance to hospital in Basingstoke where he was diagnosed as suffering from an aneurysm. He was transferred to St Peter's Hospital, Chertsey, Surrey where he died later that same evening after an operation at the age of 85.

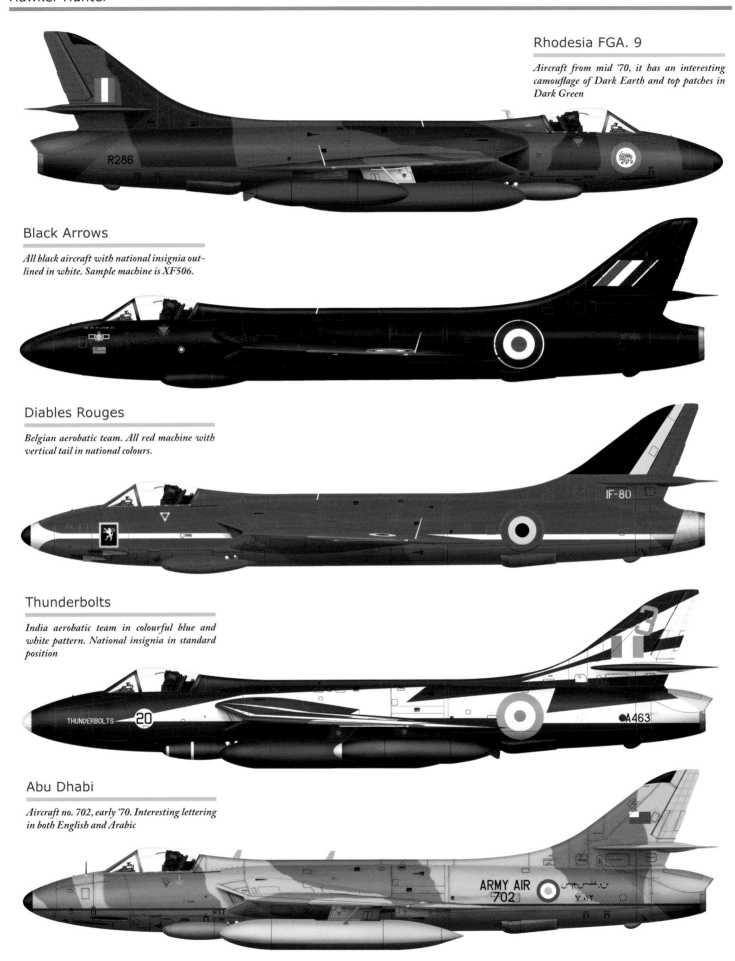

Rhodesia FGA. 9

Aircraft from mid '70, it has an interesting camouflage of Dark Earth and top patches in Dark Green

R286

Black Arrows

All black aircraft with national insignia outlined in white. Sample machine is XF506.

Diables Rouges

Belgian aerobatic team. All red machine with vertical tail in national colours.

IF-80

Thunderbolts

India aerobatic team in colourful blue and white pattern. National insignia in standard position

THUNDERBOLTS 20

A463

Abu Dhabi

Aircraft no. 702, early '70. Interesting lettering in both English and Arabic

ARMY AIR
702

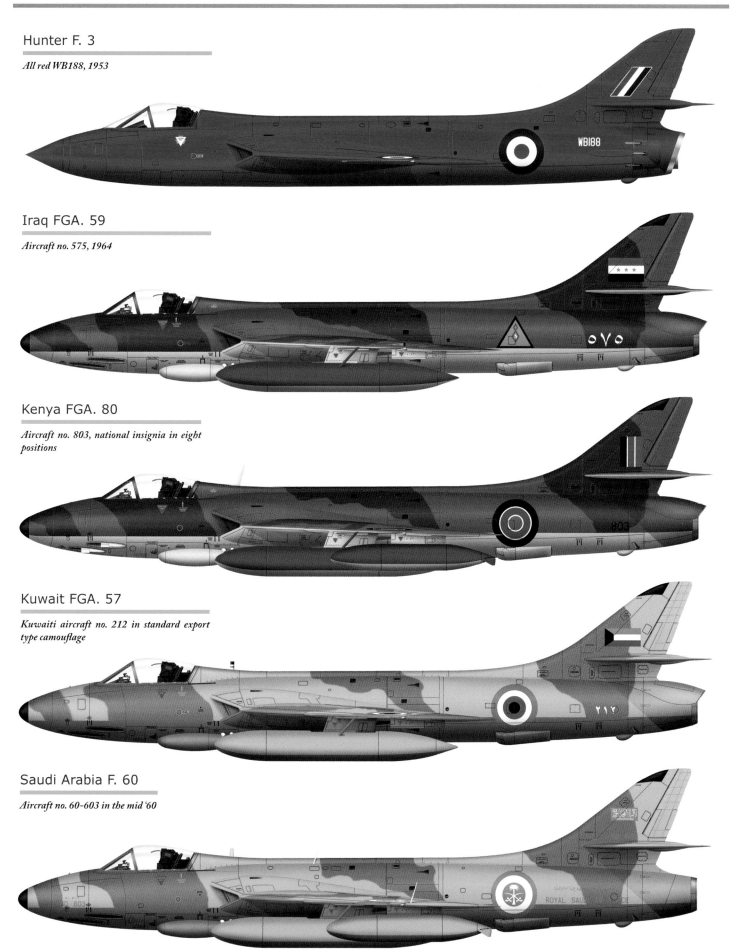

Hunter F. 3

All red WB188, 1953

Iraq FGA. 59

Aircraft no. 575, 1964

Kenya FGA. 80

Aircraft no. 803, national insignia in eight positions

Kuwait FGA. 57

Kuwaiti aircraft no. 212 in standard export type camouflage

Saudi Arabia F. 60

Aircraft no. 60-603 in the mid '60

Royal Navy

Formation start of a number of Royal Navy GA.11 Hunters. They were mainly used for weapon training, although they lacked the 30mm Aden guns.

FR. 10

XF429 was the prototype for the FR10 photo reconnaissance plane modified from the F.6.

Recce XF459

XF459 was one of the operational FR.10 photo reconnaissance Hunters. Note camera window on the side of the nose.

Early F.6

XG129 was an early production F.6 still without the characteristic 'dog-tooth' wing leading edge. It carried no operational markings, but on the nose the Hawker company logo can be seen since it was used for flight testing with underwing stores. It carries two additional drop tanks under the wings. Hunter F.6 XG128 was used for the same purposes! XG129 was later used by No. 111 Squadron for aerobatic displays in their famous Black Arrow team.

was placed in 1959 and the converted Hunters were delivered during 1960-1961. The pilots that were to fly this version got a specialized training in low level flying at no. 229 Squadron OCU (Operational Conversion Unit) at RAF Chivenor on the River Thames estuary in North Devon.

Serial numbers of the FR.10's were: WV593-WV596, XE556, XE579, XE580, XE585, XE589, XE596, XE599, XE605, XE614, XE621, XE625, XE626, XF422, XF428, XF432, XF436, XF438, XF441, XF457-XF460, XG127, XG168, XJ633, XJ694 and XJ714. XF429 served as the prototype.

After their training was completed, the pilots and their planes were assigned to No.2, No. 4 and No.79 Army-Cooperation squadron based at Gutersloh in Germany. In the summer of 1967 two No. 2 Sq. FR.12's were used for photographic missions over Gibraltar when Spain put pressure on the sovereignty of this British enclave. In 1964 FR.10's were also active at No.8 Squadron based at Khormaksar in Aden to photograph the results of earlier strike missions against rebels during the 'Rafdan' campaign. In 1967-1968 they were once more active in the Gulf area as 'watchdogs' to prevent conflicts. The Hunter FR.10 was gradually phased out from 1970 on, being replaced by Harriers.

Hunter GA Mk.11

In 1961, The British Admirality ordered 40 Hunters as single-seat weapon trainers fitted with a TACAN navigation system for use by the Fleet Air Arm. They were converted from low-hours F.4's. The four Aden guns were removed with the gun ports sealed over and ballast was installed to compensate for the changed centre of gravity. The characteristic gun bulges under the front fuselage were removed as well. They were fitted with naval radio equipment and had provision under the wings to fit rocket batteries on inboard and outboard pylons for weapon training. A small number was even fitted with cameras. The FAA Hunters received the designation GA.11 for Ground Attack. The first GA.11 was completed in April 1962. Delivery took place over 1962-1963. As true naval air-

Flight

A beautiful in-flight shot of R.A.F. Hunter F.6 XG186.

Hawker P.1083 'Super-Hunter'

With the P.1067 and later the final P.1099 Hunter Mk.6 the Royal Air Force had an excellent fighter in service. However, the Hunter was unable to reach supersonic speed in horizontal flight. It could only exceed the speed of sound for a short period if put into a dive.

Even the Hunter prototype WB188 with an up-rated jet engine fitted with an afterburner just lacked the power to go supersonic.

Camm's design team came up in 1951 with a very practical and straightforward solution. They modified a standard Hunter airframe with a new and thinner wing with a higher sweep as the Hawker P.1083. With a wing set at a sweep of 50° and fitted with the latest and most powerful Avon engine with an afterburner this new aircraft would be capable to exceed Mach 1 in level flight..

One prototype was ordered under a contract based on an existing airframe of the P.1067. R.A.F. serial was WN470. Hawker had the starboard wing already completed when the Air Ministry terminated the whole project in July 1953 for budgetary reasons.

However, this was not the final end since the partially completed airframe was used to build the Hunter F.6 prototype XF833. This aircraft made its first flight on 24 January 1954 by Neville Duke.

The P.1083 would have been capable to reach Mach 1.2 or 1235 km/h at 10,973 m with a service ceiling of 18,136 m powered by a Rolls Royce Avon RA.19R giving 5575 kg thrust at sea level and 8051 kg with full afterburner.

craft they were fitted with arrester hooks under the tail for flight deck training on airfields but it never made any deck landing.

The following GA.11's were built:
WT711-WT714, WT718, WT721, WT723, WT741, WT744, WT804-WT806, WT808-WT-810, WV256, WV257, WV267, WV374, WV308-WV382, WW654, WW659, XE668,

Landscape

A Hawker Hunter FGA.9 from no. 79 squadron flying over a typical English landscape.

Two seat

The mock-up of the earlier side-by-side front fuselage of the Hunter two-seater. The shape of final two-seater was quite different!

XE673, XE674, XE680, XE682, XE685, XE689, XE707, XE712, XE716, XE717, XF291, XF297, XF300, XF301, XF368 and XF977.

The GA.11's were used by the Royal Navy squadrons no.738 at Brawdy, no.764 at Lossiemouth and at the naval training units at Yeovilton. Some GA.11's were experimentally modified to carry Martin Bullpup and Philco Sidewinder air-to-air missiles. In 1970 there still was such a great requirement for these Hunters that the Admirality temporary acquired a small number of R.A.F. F.4's without being changed to GA.11 standards. The Hunter GA.11 remained in service until they were replaced by Hawks in 1995.

XJ615

The XJ615 Hunter two-seat prototype with a metal hood for test purposes. The aircraft carries on its nose the Hawker logo

ETPS

The XJ615 in its final form was used by the Empire Test Pilots School (E.T.P.S.).

Royal Navy T.8

XL584 was a Royal Navy Hunter T.8. We see it here in 1988 at Yeovilton. It was fitted with an arrester hook!

Kemble

An in flight shot of Hunter T.7 XL592. This one also survived; it is stored at Kemble.

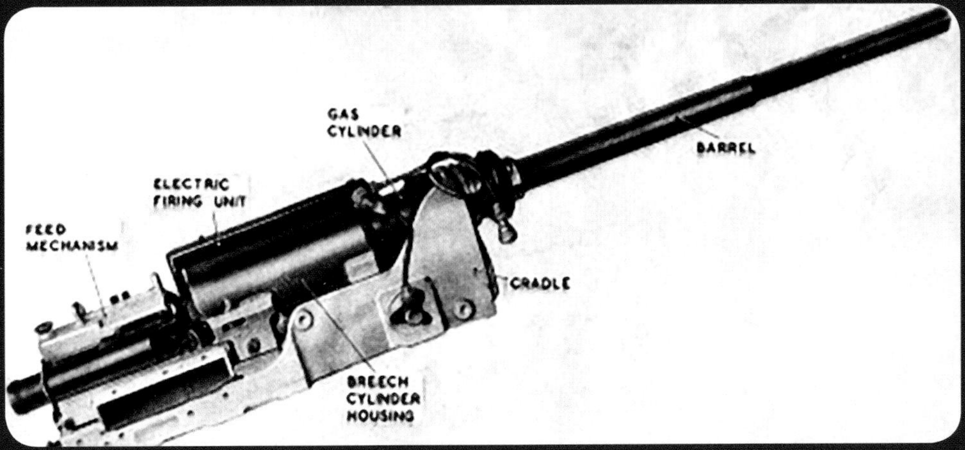

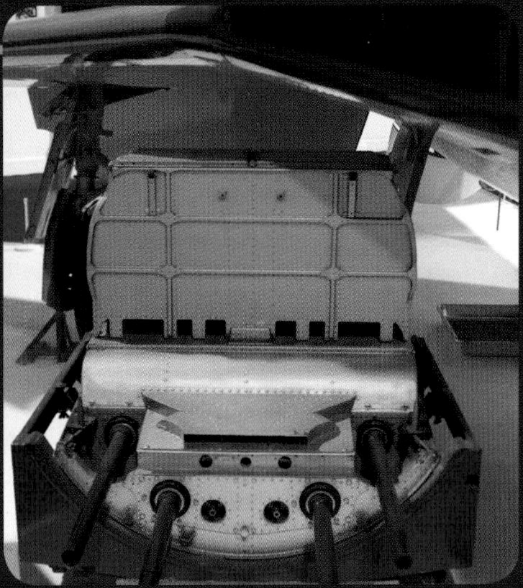

Aden 30 mm cannon

The Royal Small Arm Factory at Enfield designed a new rapid firing gun in 1946, which was to be used on fast jet aircraft as a replacement for the 20 mm Hispano-Suiza cannon. This became necessary since the time interval to deliver a fatal blow was very short in aerial combat with high speed jet planes.

The new gun had a calibre of 30 mm and was called the Aden cannon (Armament Development Enfield).

It was broadly based on the wartime German Mauser MG 213C, an experimental revolver cannon designed for the Luftwaffe that was never used operationally. The gun was fully automatic in all respects once it was loaded and the first round had been fired. The gun had to be loaded before take-off, no provision being made for loading in flight. The Aden was used as standard armament for the Hawker Hunter and the Vickers Armstrong Swift.

Technical details:

Type:	single barrel gun with rotating revolver-type feeding mechanism
Action;	recoil operation with a pneumatic cocking system powered by 26V DC
Weight:	87.1 kg (barrel weight 12.25 kg)
Length:	1.64 m (barrel length 1.08 m)
Caliber:	30 mm
Shell:	30 x 113 mm
Rate of fire:	1200-1400
Muzzle velocity:	741 m/sec

Hunter Installation

In the Hunter, the gun installation consisted of four Aden guns (two left-hand feed, two right-hand feed), carried, together with their ammunition, in a removable pre-armed armament package housed in the underside of the front fuselage. The gun barrels, which are detachable from the guns to allow removal of the package, extended forward in blast tubes below the cabin floor to apertures in the nose of the fuselage. Empty cartridge cases were ejected into the air stream through chutes extending aft from the rear of the gun package, two on each side, to apertures in the gun and radio access panels, while the belt links passed down chutes extending from the guns to collector tanks fitted beneath the fuselage. Whenever the guns were fired, the gun package was ventilated by the opening of a small electrically-operated air scoop in the gun bay starboard access panel. The guns were cocked pneumatically before flight from a ground supply. Each gun had a supply of 150 rounds.

Hunter MK.12

One Hunter two-seater, no. XE531, was ordered by the Ministry of Supply on behalf of the R.A.E. (Royal Aircraft Establishment) at Farnborough as a test aircraft. It was converted from a standard F.6 to feature a two-seat cockpit with a head-up display and a vertical nose camera. It was fitted with a Rolls Royce Avon 208 engine and delivered to the R.A.E. on March 8th 1963. It crashed at take off at Farnborough on March 17th 1982.

The two-seat Hunters

When the first Hunters entered squadron service there was no dual seat training version available and advanced fighter training had to be done with the two-seat Meteor or Vampire/ Venom. These types definitely did not have flying properties that even resembled those of the Hunter. At Hawker, Camm's design team had already expected this. Hawker had already started with the dual seat Hunter in 1953 as a private venture, with designs for both tandem and side-by-side seat arrangements. In 1954, the Air Ministry finally released a specification (Spec. T.257D) for the new trainer and Hawker submitted both designs. The side-by-side layout was selected and the construction of one prototype was ordered. It received the R.A.F. serial number XJ615. Hawker gave the new trainer the type number P.1101. XJ615 was still based on the F.4 airframe. It made its first flight on 8 July 1955 but early flight testing gave disappointing performances because of airflow disruptions around the wider cockpit. An extensive program was conducted with various cockpit hood types and streamlined fairings around the cockpit section. By mid-1956 all problems were solved with a four panel window shield and a well streamlined new fuselage section that was much more in line with the area-ruling principles. In fact the final solution was so good that the dual-seat Hunter was even slightly faster than the single seat version at the same weight! XJ615 in its final form was later used by the Empire Test Pilot School (ETPS).

A second prototype ordered by the Air Ministry was based on the F.6 (XJ627), including the more powerful Avon 203 engine. This variant was flown in 1956 with the new final hood fairings. However, the first production models had the lower rated Avon of the Mk.4 and an armament of only one Aden gun (with only one single bulge under the fuselage!). The first production model of the new Hawker trainer, designated as the Hunter T. Mk.7, made its first flight on October 11th 1957 and the type soon found its way to the various training units. The first training unit to use the Hunter T.7 was No.229 Operational Conversion Unit (OCU), receiving their first machines in 1958. The earlier

Sea Harrier inside

Royal Navy Hunter T.8 XL603 was fitted with a Hawker Sea Harrier radar installation.
It is shown here in 1985 at Fairford. It was operational until 1993 and went into private hands. During 2010, ownership of XL603 passed to Jaime Pinto and the aeroplane is currently for sale. Although not in flying condition, the airframe is complete and has an engine fitted. It is registered on the FAA registry as N419ZS.

Saudi

Side view of the Royal Saudi air force Hunter T. Mk.70 trainer 70-616.

G-APUX

G-APUX was the well-known Hawker Siddeley owned demonstrator. It was later shipped to the Middle East, but finally ended up in Chile as Hunter T. Mk. 72.

T.7's still had the old F.4 wing, but on later models this was soon replaced by the F.6 wing with its extended leading edge. The T.7 was much in demand at the operational squadrons and many had a two-seater added to their strength for instrument flying training and for keeping pilots current on their jet rating. Hunter T.7's remained in service until they were replaced by the Hawk.

With the Hunter GA.11 in its inventory, the Royal Navy was interested in a dual seat Hunter too. Here it was known as the T. Mk.8. Essentially it was based on the T.7, but it also had the standard arrester hook under the tail just like the GA.11 had. Further it was fitted with the same TACAN navigation system as the GA.11. The prototype of the T.8, WW664, flew for the first time on March 3rd 1958. It was completed using a damaged but repaired F.4 airframe. The first newly built T.8, the XL580 flew on May 30th of the same year. A total of 33 T.8's was delivered to the Royal Navy; ten newly built and the others converted from existing airframes. They all saw extensive service at F.A.A. squadrons and were used for may years.

The two seat T.7 was further exported to various countries abroad where they also served for many years. There still are a number of two-seat Hunters airworthy in the airshow circuit.

License production

Fokker in the Netherlands together with SAB-CA and Avions Fairey in Belgium, license built

KLU

Dutch Hunter T.7 N-301 in flight

a total of 462 Hunters F.4 and F.6 for the Dutch and Belgian air forces.

The first Hunter F.4 for Fokker was a pattern machine carrying the military registration N-1. It was ferried from the U.K. to Schiphol on March 3rd 1955, flown by Fokker chief test pilot Gerben Sonderman. It went into service at the Koninklijke Luchtmacht (KLu) as N-201. Hawker supplied another five to Fokker as complete unassembled packages before the final production line was started up. Fokker built 96 Hunter Mk.4's for the KLu with registration numbers N-101 to N-196. The last KLu F.4 Hunter was delivered to 325 Sq. at Leeuwarden airbase on November 1st 1957. Another 12 were supplied to the Belgian Air Force, with registrations ID-101 to ID-104, ID-109, ID-110, ID-115, ID116, ID-121, ID-122, ID-127 and ID-128. Further 36 complete packages were supplied for final assembly in Belgium by SABCA and Fairey with registration nos.: ID-105 to ID-108, ID-111 to ID114, ID-117 to ID-120, ID-123 to ID-126 and ID-129 to ID-148. SABCA and Avions Fairey produced another 64 F.4's for the Belgian Air Force with registration nos. ID-1 to ID-64.

Total Hunter F.4 production in the Netherlands and Belgium, including the pattern machine, was 208.

The first Fokker built KLu F.6 Hunter was delivered on 17 October 1957 as the N-201. Fokker produced 93 F.6 Hunters for the KLu, with registration Nos. N-201 to N-293. The last Hunter F.6 was delivered on April 17th 1959.

For Belgium, Fokker delivered 92 complete F.6 packages for assembly by SABCA and Avions Fairey. Another 52 were built completely by SABCA and Avions Fairey.

Total F.6 production in the Netherlands and Belgium was 237.

When the F.4's and F.6's were phased out, many went back to Hawker for overhaul and conversion into export machines.

Hunters for everyone!

With the Hunter being successfully introduced at the Royal Air Force, it became a very important export article for Great Britain. As a successor for ageing Gloster Meteors contracts were closed with aircraft manufacturers in Belgium and the Netherlands for license production. Hunters also found their way to other European countries like Denmark, Sweden and Switzerland. This was however not the end! Hunters were also exported to Asian, Middle-East, African and South American countries. These were in most cases no newly manufactured aircraft, but refurbished Hunters purchased back by Hawker and adapted to the customer's specifications! Hawker rightfully concluded there was a big world-wide market for the Hunter at a stage when at the R.A.F., and in other European air forces too, they were already decommissioned from operational service!

European users

Belgium- The Belgian air force had in total 113 Hunter F.4's and 144 F.6's in its inventory, all built under license by Fokker or SABCA/Avions Fairey. The Belgian air force never had any Hunter trainers! The Belgian air force even had an aerobatic display group flying all-red Hunters as the 'Red Devils' ('Diables Rouges').

Denmark- In 1954, Denmark ordered thirty Hunters designated as F.Mk.51. They were based on the Hunter F.4. The registration nos. of the Danish MK.51's were E-401 to E-430, although at a later stage the 'E' was omitted. The newly built Mk.51's were delivered in 1956 and served at EKS-724. First flight of E-401 was made on December 15th 1955 by Hawker test pilot David Lockspeiser. After phasing out, a number were repurchased by Hawker Siddeley for other costumers and some of the Danish Hunters still exist.

For training of their Hunter pilots Denmark purchased two newly built T. Mk.53 trainers, based on the T.7. They carried the registrations E-271 and E-272 and were supplied in

Technical details

	Hunter F.4	Hunter F.6
Power plant:	Rolls-Royce Avon 121	Rolls-Royce Avon 207
-thrust:	7500 lbf -3402 kgf	10,145 lbf -4600 kgf
Dimensions:		
-wingspan	10.26 m	10.26 m
-length	14.00 m	14.00 m
-height	4.01 m	4.01 m
-wing area	32.42 m²	32.42 m²
Weights:		
-empty	5690 kg	5788 kg
-maximum take off	7757 kg	8050 kg
Performances:		
-Max. speed	1094 km/h at s.l.	1117 km/h at s.l.
-Service ceiling	14,874 m	15,997 m
-Range (combat)	715 km	715 km

Armament: 4× 30 mm (1.18 in) Aden revolver cannons in a removable gun pack with 150 rounds per gun (all versions except GA.11; T.7 and T.8 two-seaters only had one Aden cannon). The Hunter was fitted with an Ekco radar gun sight with the radar dish fitted in a small radome in the point of the fuselage nose.
The FGA.9 ground attack version and its related export versions had standard four hard point under the wings for various loads ranging from extra fuel tanks to bombs and rockets. Some Hunters were converted to carry underwing missiles like Philco Sidewinder and the Maverick.

From the Hunter FGA Mk. 58 manual....

These extracts from the official technical manual of the Hunter FGA Mk.58 give very clear details on the construction of this Hunter type for Qatar. It was essentially similar to the Hunter FGA.9.

The Hunter was very sturdy constructed using the basic techniques of the mid-fifties. It still was an aircraft that had to be flown by the pilot instead of by a central board computer! With hydraulically powered controls it was easy to fly without much limitations. Tight low and high speed turns, standard aerobatic manoeuvres, stalls and spins could be all safely made.

Structurally the Hunter was a very strong aircraft and that may explain why so many Hunters saw a second life after being phased out in Europe.

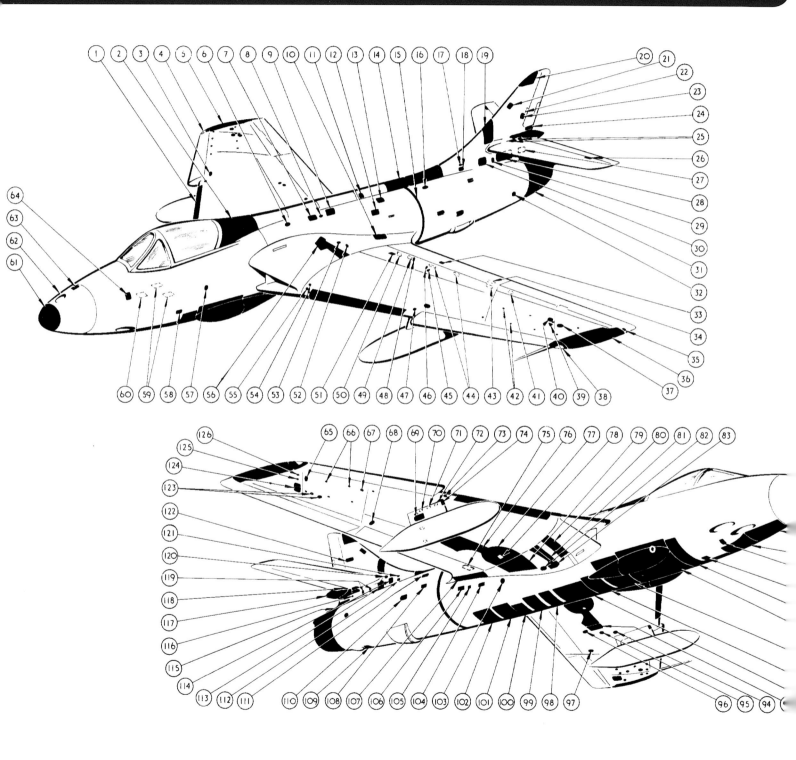

F.S./6

1	Detachable wing nosing	
2	Flying controls & cabin pressurizing	
3	General access	
4	Navigation light window	
5	Detachable wing tip	
6	Fuel level switch	
7	Fuel vent connections	
8	Starter equipment	
9	Cold air unit	
10	Rear spar pin joint & flying controls	
11	Fuel system external air connection	
12	Front engine mounting	
13	Air supply	
14	Flying controls	
15	Transport joint	
16	Fuel level switch	
17	Flying controls pivot bolt	
18	Flying controls lever	
19	Flying controls	
20	Detachable tip	
21	Rudder controls	
22	Rudder trim tab actuator	
23	Rudder controls	
24	Fin detachable portion	
25	Tail plane hinge & general access	
26	Elevator outer hinge	
27	Flying tail switch linkage	
28	Selector valve & elevator power control	
29	Tail plane actuator	
30	Hydraulic accumulator charging valve and electrics	
31	Detachable tail cone	
32	Jet pipe rear mounting	
33	Flap jack	
34	Electrical	
35	Aileron outer hinge	
36	Detachable wing tip	
37	General access	
38	Navigation light window	
39	Stores crutching	
40	General access	
41	Aileron trim tab actuator	
42	Rocket projectile mounting	

17866— 18736

KEY TO FIG. 3 (ACCESS PANELS)

43	Aileron controls	
44	Aileron controls	86
45	Flap jack anchorage	87
46	Flap jack greaser	88
47	Stores crutching	89
48	Flap synchronizing jack bleeding	90
49	Aileron controls	91
50	Drum switch, flap control	92
51	Undercarriage jack attachment bolt	93
52	Manual undercarriage catch release	94
53	Fuel connection	95
54	Slinging socket	96
55	Wing pin joint	97
56	Main spar pin joint	98
57	External emergency hood release	99
58	Pilot's footstep	100
59	Flying controls	101
60	Control column mechanism	102
61	General access	103
62	Detachable nose piece	104
63	Camera servicing	
64	General access	105
65	General access	106
66	Rocket projectile mountings	107
67	Fuel pipes	108
68	Aileron controls	109
69	General access	110
70	Electrical	111
71	Fuel & air pipes	112
72	Stores crutching	113
73	General access	114
74	General access	115
75	Hydraulic reservoir	116
76	Main undercarriage leg fairing	117
77	Main undercarriage wheel fairing	118
78	Wheel brake	119
79	Main undercarriage door	120
80	Fuel tank float switch	121
81	Fuel transfer pipe & pressure relief valve	122
82	Fuel & hydraulic pipes	123
83	Main spar pin joint	124
84	Nose undercarriage front door	125
85	Nose undercarriage rear door	126

RESTRICTED

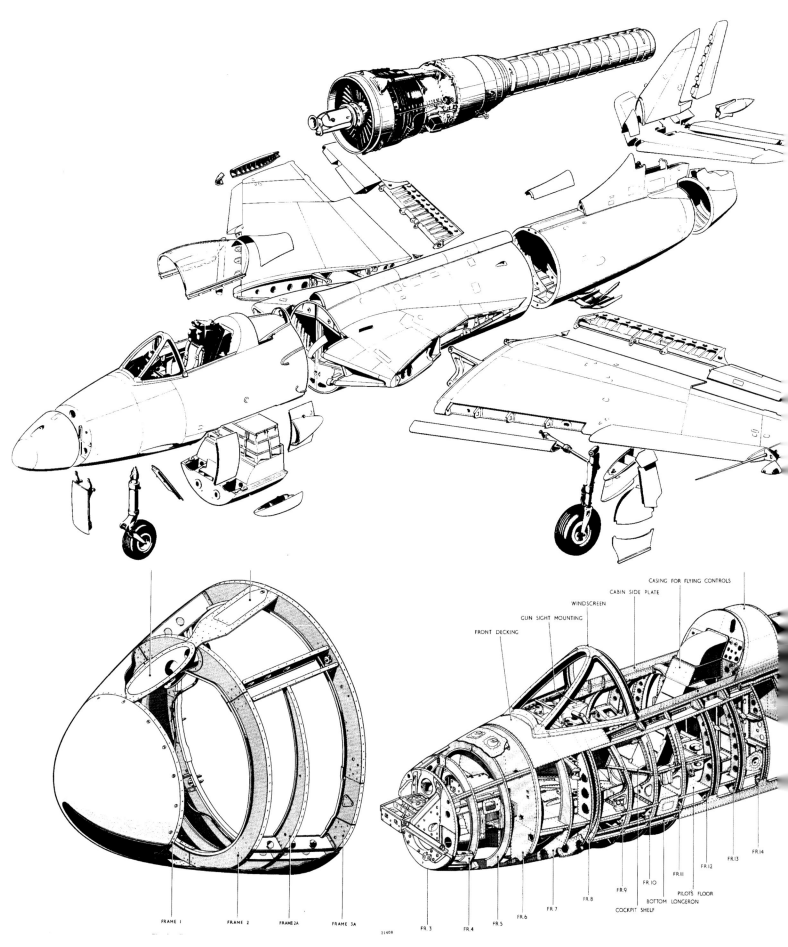

FRAME 1   FRAME 2   FRAME 2A   FRAME 3A   21408   FR.3   FR.4   FR.5   FR.6   FR.7   FR.8   FR.9   FR.10   FR.11   FR.12   FR.13   FR.14

COCKPIT SHELF   BOTTOM LONGERON   PILOTS FLOOR

FRONT DECKING   GUN SIGHT MOUNTING   WINDSCREEN   CABIN SIDE PLATE   CASING FOR FLYING CONTROLS

Fig. 2 Front fuselage

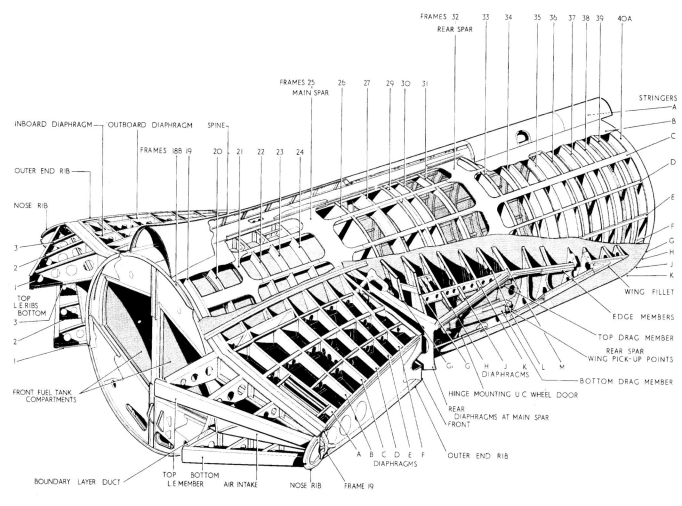

FRAMES 32
REAR SPAR

33 34 35 36 37 38 39 40A

STRINGERS
A
B
C
D
E
F
G
H
J
K

FRAMES 25
MAIN SPAR

26 27 29 30 31

INBOARD DIAPHRAGM   OUTBOARD DIAPHRAGM   SPINE

FRAMES 18B 19   20 21 22 23 24

OUTER END RIB

NOSE RIB

3
2
1
TOP
L E RIBS
BOTTOM
3
2
1

WING FILLET

EDGE MEMBERS

TOP DRAG MEMBER

REAR SPAR
WING PICK-UP POINTS

BOTTOM DRAG MEMBER

G G H J K L M
DIAPHRAGMS

HINGE MOUNTING U C WHEEL DOOR

REAR
DIAPHRAGMS AT MAIN SPAR
FRONT

FRONT FUEL TANK
COMPARTMENTS

A B C D E F
DIAPHRAGMS

OUTER END RIB

BOUNDARY LAYER DUCT

TOP   BOTTOM
L E MEMBER   AIR INTAKE

NOSE RIB

FRAME 19

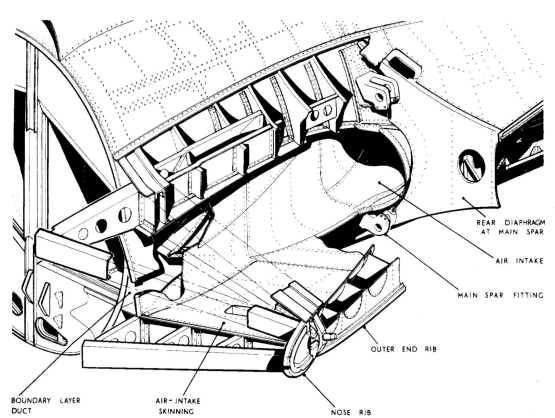

P LONGERON

FR.18A
FR.17B

REAR DIAPHRAGM
AT MAIN SPAR

AIR INTAKE

MAIN SPAR FITTING

OUTER END RIB

BOUNDARY LAYER
DUCT

AIR-INTAKE
SKINNING

NOSE RIB

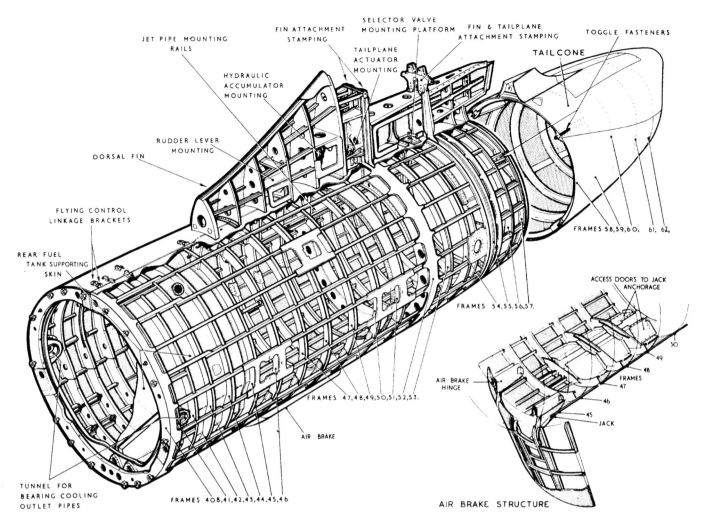

JET PIPE MOUNTING RAILS

FIN ATTACHMENT STAMPING

SELECTOR VALVE MOUNTING PLATFORM

FIN & TAILPLANE ATTACHMENT STAMPING

TOGGLE FASTENERS

TAILCONE

HYDRAULIC ACCUMULATOR MOUNTING

TAILPLANE ACTUATOR MOUNTING

RUDDER LEVER MOUNTING

DORSAL FIN

FLYING CONTROL LINKAGE BRACKETS

REAR FUEL TANK SUPPORTING SKIN

FRAMES 58, 59, 60, 61, 62,

FRAMES 54, 55, 56, 57.

ACCESS DOORS TO JACK ANCHORAGE

AIR BRAKE HINGE

50

49

48

FRAMES 47

46

45

JACK

FRAMES 47, 48, 49, 50, 51, 52, 53.

AIR BRAKE

TUNNEL FOR BEARING COOLING OUTLET PIPES

FRAMES 40B, 41, 42, 43, 44, 45, 46

AIR BRAKE STRUCTURE

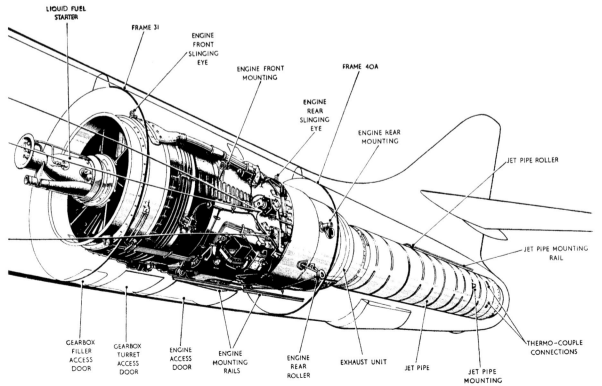

LIQUID FUEL STARTER

FRAME 31

ENGINE FRONT SLINGING EYE

ENGINE FRONT MOUNTING

FRAME 40A

ENGINE REAR SLINGING EYE

ENGINE REAR MOUNTING

JET PIPE ROLLER

JET PIPE MOUNTING RAIL

GEARBOX FILLER ACCESS DOOR

GEARBOX TURRET ACCESS DOOR

ENGINE ACCESS DOOR

ENGINE MOUNTING RAILS

ENGINE REAR ROLLER

EXHAUST UNIT

JET PIPE

JET PIPE MOUNTING

THERMO-COUPLE CONNECTIONS

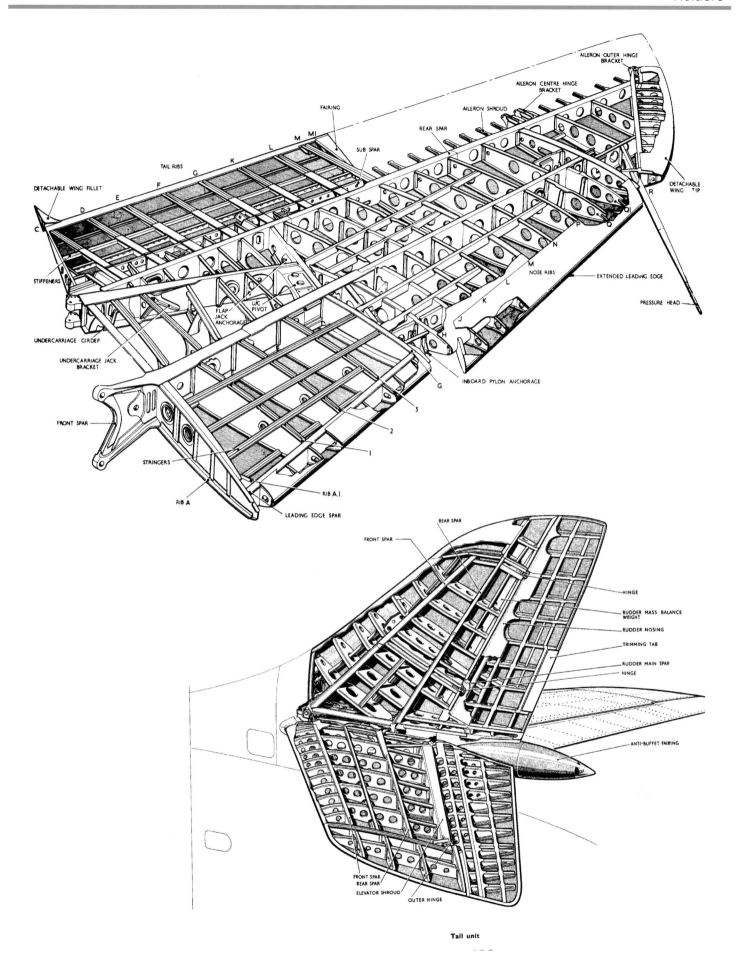

AILERON OUTER HINGE BRACKET

AILERON CENTRE HINGE BRACKET

AILERON SHROUD

FAIRING

REAR SPAR

SUB SPAR

DETACHABLE WING TIP

TAIL RIBS

DETACHABLE WING FILLET

M M1
L
K
G
F
E
D
C

R

Q1
Q
P

N

M

STIFFENERS

EXTENDED LEADING EDGE

NOSE RIBS

L

K

PRESSURE HEAD

UNDERCARRIAGE GIRDER

FLAP JACK ANCHORAGE

U/C PIVOT

J

H

UNDERCARRIAGE JACK BRACKET

INBOARD PYLON ANCHORAGE

G

3

FRONT SPAR

2

STRINGERS

I

RIB A.1.

RIB A

LEADING EDGE SPAR

REAR SPAR

FRONT SPAR

HINGE

RUDDER MASS BALANCE WEIGHT

RUDDER NOSING

TRIMMING TAB

RUDDER MAIN SPAR

HINGE

ANTI-BUFFET FAIRING

FRONT SPAR
REAR SPAR
ELEVATOR SHROUD

OUTER HINGE

Tail unit

Production list

Production list of all Hunter types is as follows:

P.1067 prototypes:	WB188, WB195, WB202 (3)	
Hunter F.1:	WT555-WT595 (41)	
	WT611-WT660 (50)	
	WT679-WT700 (22)	
	WW599-WW610 (12)	
	WW632-WW645 (14)	all F.1's built by Hawker
Hunter F.2:	WN888-WN921 (34)	built by Armstrong Whitworth (A.S. Sapphire engine)
	WN943-WN953 (11)	
Hunter Mk.3:	was WB188 prototype modified for R.R. Avon with reheat	
Hunter F.4:	WT701-WT723, WT734-WT780, WT795-WT811 (85)	
	WV253-WV281, WV314-WV334,WV363-WV412 (100)	
	WW589-WW591 (3)	
	WW646-WW665 (20)	
	XE657-XE689, XE702-XE718, XF289-XF324, XF357-XF370 (100)	
	XF932-XF953, XF967-XF999, XG341-XG342 (57)	
Hunter F.5:	WN954-WN992, WP101-WP150, WP179-WP194 (A.S. Sapphire engine) (105)	
Hunter P.1083	WN470 with 45° wing; cancelled before completion	
Hunter P.1099:	XF833 Mk. 6 prototype built from incomplete P.1083 (1)	
Hunter F.6:	WW592-WW598 (7) pilot production batch	
	XE526-XE561, XE579-XE628, XE643-XE656 (100)	
	XG127-XG137, XG169-XG172, XG185-XG211, XG225-XG239,	
	XG25XG274, XG289-XG298 (110) built by Armstrong Whitworth	
	XJ632-XJ656, XJ676-XJ695, XJ712-XJ718 (45)	
	XK136-XK176, XK213-XK224 (53), 100 cancelled	
Hunter P.1101:	XJ615 and XJ627 prototypes for two-seat T.7 trainer (2)	
Hunter T.7:	XL563-XL579, XL583, XL586-XL587, XL587, XL591-XL597, XL600,	
	XL601, XL605, XL609-XL623 (45)	
Hunter T.7:	Dutch order; KLu registrations N-301-N310, N311-N320 (20)	
Hunter T.8:	XL580-XL582, XL584, XL585, XL598, XL599, XL602-XL604 (10)	
	Other T.8 trainers were converted from existing F.4 and F.6 airframes.	
Hunter FGA.9:	128 converted from F.6	
Hunter FR.10:	32 converted from F.6 + one prototype	
Hunter Mk.11:	40 converted from F.4 for use at the Royal Navy without the Aden guns. They had ar rester hooks, but were never used on aircraft carriers.	
Hunter Mk.12:	conversion of T.6 XE531 for Royal Aircraft Establishment at Bedford	
Hunter Mk.50:	Swedish order for Mk.4 nos.34001-340120 (120)	
Hunter Mk.51	Danish order for Mk.4 nos. 401-430 (30)	
Hunter Mk.52:	Peruvian order for 16 refurbished F. Mk.4's	
Hunter Mk.53:	Danish order for T.7 trainer nos.37-271 and 35-272 (2)	
Hunter Mk.56:	Indian order for 160 Mk..6; 32 from R.A.F. stock. Remainder newly built, nos.BA249-BA360 (128)	
Hunter Mk.57:	Kuwaiti order, 4 delivered from ex-Belgian Mk.6 as FGA.9	
Hunter Mk.58:	Swiss order for 12 converted from existing stock and 88 newly built as J-4001 to J-4100. Based on F.6.	
Hunter Mk.59:	Iraqi order for 46 Mk.6 converted from existing stock including refurbished ex-Dutch	
Hunter Mk.60	Order for Saudi Arabia, Jordan and Oman; converted from existing stock	
Hunter Mk.62:	Mk.4 WT706 converted to two-seater for Peru	
Hunter Mk.66:	two-seaters for India, Jordan and Lebanon; converted from existing stock	
Hunter Mk.66B	two-seaters for Jordan and Oman	
Hunter Mk.66C	two-seater for Lebanon	
Hunter Mk.66D/E	two-seaters for India	
Hunter Mk.67	two-seaters for Kuwait and Oman	
Hawker Mk.68	two-seaters for Switzerland	
Hunter Mk.69	two-seaters for Iraq	

Hunter FGA Mk.70   for Lebanon, based on FGA.9
Hunter T.Mk.70   two-seater for Saudi Arabia and Jordan
Hunter FGA. Mk.71   for Chile; also as Mk.70A
Hunter T. Mk.72   two-seater for Chile
Hunter FGA Mk. 73   for Jordan and Oman; also as Mk.72A and B
Hunter FGA Mk.74   for Singapore; also as MK.74A and B
Hunter T. Mk.75/75A two-seaters for Singapore
Hunter FGA Mk.76   for Abu Dhabi and Somalia; also as Mk.76A
Hunter T.Mk. 77   two-seater for Abu Dhabi and Somalia
Hunter FGA. Mk.78   for Qatar
Hunter T. Mk.78   two-seater for Qatar
Hunter FGA Mk.80   for Kenya and Zimbabwe
Hunter T. Mk.81   two-seater for Kenya and Zimbabwe
P.1109A and B: converted from F.6 for experiments with Firestreak air-to-air missiles. Two built with
                     extended nose; WW549 andXF378
P.1128 Huntsman: projected six-seat military executive jet based on Hunter; not built
Fokker production: Mk.4 N-101 to N-196 (96); N-101 supplied by Hawker as pattern (Dutch only)
                     machine'N-1' and Mk.6 N-201 to N-293 (93)
Belgian supply: Mk.4 ID-1 to ID 64 (64) and ID101 to ID148 (48) and Mk.6 IF-1 to IF-144 (144) built
                     by Fokker, SABCA and Avions Fairey

Hawker produced the Hunter not only at the Kinston-upon-Thames plant, but also at their production facility
at Blackpool. Also Armstrong Whitworth at Coventry produced a large number of Hunters.
Production numbers at the various locations were:

Hawker Kinston-upon-Thames: 972
Hawker Blackpool: 299
Armstrong Whitworth Coventry: 269

Belgian Hunter

A Belgian Hunter F. Mk. 4; no. ID44

No. IF-61 was a Belgian Hunter F.6

Denmark

Hunter F. Mk.51 No. 474 in a later three-tone camouflage scheme with the original suffix 'E' removed.

Hawker Hunter F.6 N-257 from 325 sq. Soesterberg. It was later purchased back by Hawker Siddeley and converted as a T. Mk. 57 two-seater with registration no. 219.

India

Indian air force pilot in front of a Hunter F. Mk.56. During armed conflicts with Pakistan it was found to be on quite even terms with the Pakistan F-86E Sabre fighters. The Sabre was in some points superior since the Hunter lost speed in tight turns during combat manoeuvres, but the Hunter's armament of 30 mm cannons was definitely better that the Sabre's .50 machine guns!

Display team

A mixed group of Indian air force Hunters with a two-seat T. Mk.66 at the foreground.

1958. A further two were converted from ex-Dutch Hunters T.7 N-301 and N-307. They had the Danish military registrations E-273 and E-274. After phasing out they were re-purchased by Hawker Siddeley and they finally ended up at the Duxford aviation museum.

Netherlands- Except for the Hunters F.4 and F.6 built by Fokker, the KLu also had twenty Hunter T.7 trainers. They were ordered directly at Hawker and were newly built. They carried the registrations N-301 to N-320. They were delivered during the period July 1958-April 1959.

The last KLu T.7, N-320, went to the NLR (National air and space laboratories) as a general fast flying calibration aircraft to replace the earlier Fokker S.14 Machtrainer PH-XIV. It flew with the civil registration PH-NLH and was used over the period 1966-1980, making a total of 532 flying hours.

Sweden- The Swedish air force (Flygvappen) ordered 120 Hunter F.Mk.50's in June 1954. The F.50 was more or less equivalent to the Mk.4. Registration numbers used were 34001 to 34120. The Swedish Hunter was intended as a 'stop-gap' between the SAAB J 29 Tunan and the supersonic SAAB J 35 Draken. Especially

Swiss F. 58

Swiss machine in winter camouflage. This type of camouflage did not have a wide service use

Hunter T.Mk.12 XE531

Used as a general test aircraft by the Royal Aircraft Establishment (R.A.E.) at Farnborough.

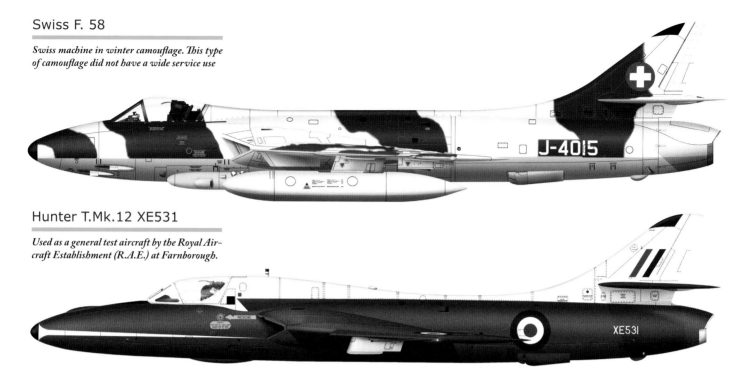

the practical service ceiling of the Tunan was regarded as being too low to intercept high-flying enemy bombers. The first aircraft was delivered in August 1955 to the Flygvappen, where it was designated as the J 34. The Mk.50's were used by F.8 and F.10 Wings for the direct air defence of the capital of Stockholm. One of the most important modifications was that the Mk.50 was made suitable to be armed with two underwing Philco Sidewinder air-to-air missiles. Sweden also had an aerobatic display team using five Hunters that became known as the 'Acro Hunters'. With the arrival of the first J 35 Drakens the Hunters were re-assigned to the less prominent air wings F.9 at Gothenburg and F.10 at Ängelholm and finally phased out. The last Hunters were phased out in the late sixties. Again, a number of Swedish Hunters were repurchased by Hawker Sidddeley for refurbishing and sale abroad.

Switzerland- Just like Sweden, Switzerland had a tradition of military neutrality and both countries were not involved in both World Wars! However, neutrality had its price and it implied that the armed forces had to be well equipped with the most modern weapons to defend their territory. Sweden successfully set up a national aviation industry that was, especially after the Second World War. fully capable to produce aircraft types that could fully meet themselves with the most modern contemporary types. Also Switzerland tried to accomplish this, but with less success. After the war Switzerland tried to produce its own jet fighter, the E.F.W. N-20 Aiguillon. It was a technical failure and finally Switzerland purchased the British De Havilland Venom jet. When these had to be replaced they again tried

to develop a new jet fighter as the F.F.A. P-16. The P-16 was no technical success either and after two prototypes crashed the project was ended. Though......the designer of the P-16 went to the United States and developed the P-16 into something totally new at that time: a jet powered business aircraft. Even today it is still well known as the Learjet!!

With the P-16 being cancelled, Switzerland again decided 'to buy British'. In 1958 this resulted in an order for 100 Hunters, based on the F.6. Hawker designated these aircraft as the F.Mk.58. The first twelve were converted from ex-R.A.F. machines, but the other 88 were newly built. These aircraft received the Swiss air force registration nos. J-4001 to J-4100. They were used until the nineties when they were replaced by the American F-5E. The Swiss air force also used the Hunter for their well-known aerobatic display team, the 'Patrouille de Suisse'.

Switzerland also ordered thirty two-seat Hunters. These were not newly built, but refurbished and converted from existing stock. Based on the Hunter T.7, Hawker gave these aircraft the designation Mk.58A. At the Swiss air force they carried the registration numbers J-4131 to J-4162. Two of these Mk.58A Hunters were the former Dutch T.7 Hunters N-313 and N-318.

Middle-East users

Abu Dhabi- The air force of Abu Dhabi purchased seven Hunter single seat fighters, designated as FGA Mk.76. They were based on the FGA.9 and converted from R.A.F. Mk.4 stock aircraft WV389, WV402, XE589, XF362,

Kenya

A Kenyan Hunter FGA. Mk.80 with registration number 805.

Qatar

Rare shot of a Hunter FGA Mk.78 from Qatar. It shows no. QA-11; former Dutch KLu Hunter F.6 N-219.

Somalia FGA. 76

Former Abu Dhabi aircraft. It retained its original camouflage with only the national insignia changed

Oman F. 73

Oman had a very interesting camouflaged aircraft with two shades of dark grey in wrap-around pattern.

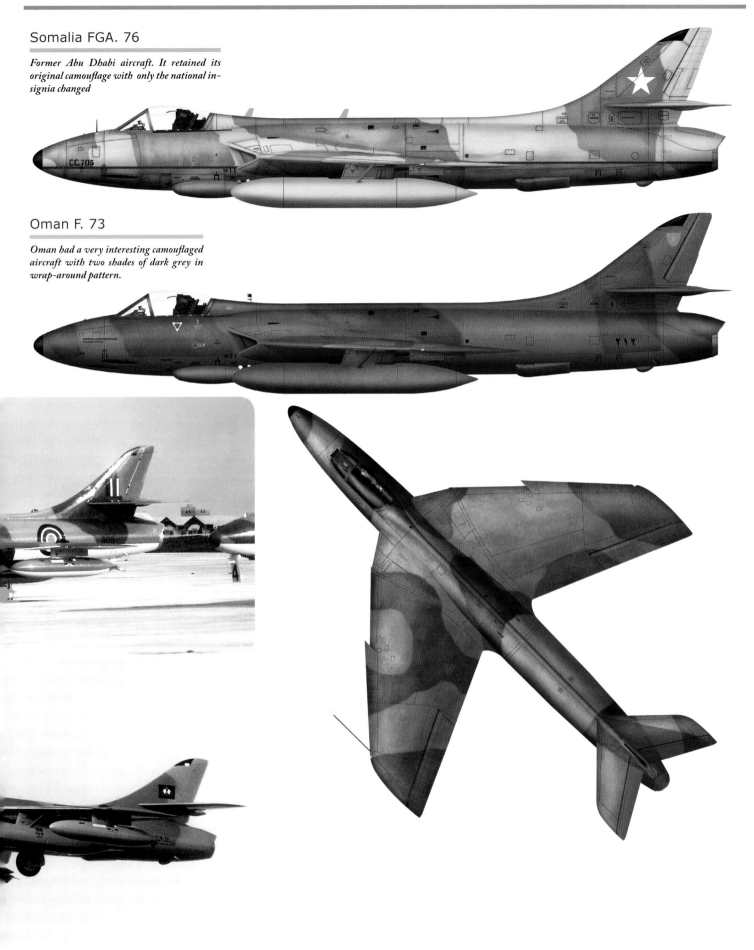

Rhodesia

An unmarked and anonymous Rhodesian air force Hunter FGA.9

Saudi

Saudi Arabian Hunter F. Mk. 60 no.60-602

XF367, XF935 and XG341. In Abu Dhabi they were renumbered as 701 to 707. Another three Hunters, designated as the Mk.76A, were modified to FR.10 standards with the same camera equipment. They were former R.A.F. Hunters Mk.4 nos. WV400, XF971 and Hunter Mk.6 no. XF971. Abu Dhabi further ordered two T.7 trainers, designated as T. Mk.77. They were ex-Dutch KLu Hunters N-301 and N-312 refurbished and brought up to the latest standard by Hawker Siddeley. All these machines were delivered in 1970-1971.

Iraq- Between 1964 and 1975, both Britain and France delivered significant quantities of arms, including Hunters, to Iraq. The Hunters were far more effective in fighting guerrilla activity than the Russian MiG's then operated by Iraq. Iraq placed a first conversion order for 24 Hunters FGA Mk.59, converted from Ex-Belgian and Ex-Dutch Hunters F.6. They were brought up to full FGA.9 standards and were delivered in 1964-1965. They received the military serial numbers 570 to 587 and 628 to 633. A second order was placed for 18 aircraft, again refurbished ex-Belgian and ex-Dutch Hunters brought up to FGA.9 standard. This batch was delivered in 1965-1967 and designated as FGA Mk.59A. Registration numbers were 657 to 661 and 690 to 702.

Iraq also purchased in two batches five Hunter T. Mk.69 two seat trainers, being refurbished

and converted ex-Belgian Mk.6 single-seaters. They received registration nos. 567 to 569 and 626 to 627.

Jordan- Jordan received, as a present from Saudi Arabia, three Hunters F. Mk.60 in 1967. They were converted from ex-RAF F.6 Hunters.

Furthermore, Jordan received in total ten refurbished Hunters, based on the FGA.9, under designation FGA Mk.73, Mk.73A and Mk.73B. They were supplied in 1969, 1970 and 1971, receiving the registration nos. 814, 828 to 831, 832, 840, 841, 850 and 851.

Jordan leased the Hawker Siddeley Hunter two-seat demonstrator G-APUX for some time in the mid-sixties as the T. Mk.66A, but the plane finally ended up in 1975 in Chile as the T. Mk.72. Jordan purchased three additional Hunter two-seaters as the T. Mk.66B. One was newly built to replace G-APUX; the two others were converted from Ex-Dutch Mk.6 fighters

Lebanon- Already in 1958 the Lebanese air force received six ex-R.A.F. Hunter F.6 under a U.S funded so-called off-shore procurement. The Lebanese air force purchased another four Hunters FGA Mk.70 refurbished from ex-Belgian Hunters F.6 with registration nos. L176 to L179. They were delivered between September 1965 and September 1966. Lebanon leased the Hunter T. Mk.66 (G-APUX)

Jordan

A pair of ex-Saudi Arabian air force Hunters F. Mk. 60 of the Jordanian air force

Singapore

Hunter FGA Mk. 74 from the air force of Singapore.

for some time before it was used by the Jordan air force, The Lebanese air force further operated three T. Mk Mk.66A Hunter trainers refurbished from ex-Belgian Mk.6 Hunters. They were delivered between November 1965 and July 1966 and carried the registration numbers L280 and L282. The Hunters were used in military actions against Israel and a number were destroyed. During the civil war in Lebanon the Hunters were stored, but in November 2008 they were put back into service, fifty years after its introduction at the R.A.F.! Even in the summer of 2010 they were still in use!

Kuwait- Four ex-Belgian F.6 Hunters were converted to FGA 9 specifications as the FGA Mk. 57. They were delivered over 1965-1966 receiving the registration numbers 212 to 215.

Kuwait also ordered five Hunter two-seat trainers as the T. Mk.67. Two were converted ex-Belgian F.6 Hunters. The other three were ex-Dutch F.6 Hunters N-257 and N-282 and ex-R.A.F. Hunter F.6 XE530. Kuwaiti registration numbers were 210, 211, 218, 219 and 220.

Oman- In 1975, the Sultanate of Oman received 31 Hunters FGA Mk.73 as a gift from King Hussein of Jordan and from Kuwait. Sixteen were put into service, operated mostly by British pilots. The other Hunters were stored for spares supply. In 1980 the Hunters still remaining were

fitted with underwing pylons to carry two Philco Sidewinder missiles. Oman also had five T. Mk 66 and T. Mk 67 trainers in service, being ex-Kuwaiti and ex-Jordan machines. The Hunters remained in service until 1993.

Qatar- Qatar purchased three refurbished and converted ex-Dutch Hunters F.6 (N-219, N-222 and N-268) as FGA Mk.78 receiving registrations QA-10 to QA-12.

A Hunter T. Mk.78 two-seater for Qatar for Qatar was refurbished from the ex-Dutch Hunter T.7 N-316.

Saudi Arabia- Six Hunters F. Mk.60 were supplied in 1966 to Saudi Arabia to ease the transition to the supersonic Lightning. They were refurbished F.6's and received the registrations 60-601 to 60-604. One was lost, the remaining five were donated to Jordan. The Saudi air force also had two two-seat Hunters T. Mk. 70 nos.70-616 and 70-617 (ex RAF T.7's XL605 and XL620). They were delivered in 1966 and used until 1974.

Far East users

India- As early as 1954 India was already making inquiries for Hunter purchase for the Indian air force. This took place at a time the neighbouring country Pakistan was in negotiation with the

U.S.A. to obtain the North American F-86 Sabre jet fighter. India finally ordered the Hunter F. Mk.6. Hawker designated the Indian F.6 as the Hunter Mk.54 and with this order India was one of the biggest Hunter customers! In September 1957 it was in fact the first overseas order placed! The first 32 were already produced for a British Military of Supply contract, but eventually the machines were diverted to India as the Mk.56. The remaining 128 Hunters Mk.56 were newly built at Hawker.

A second production order followed some ten years later for 36 Hunters Mk.56A, built to FGA.9 standards as ground attack fighters. They were ex-Belgian and ex-Dutch Hunters Mk.6.

The Indian air force also acquired a number of two-seat Hunters.

The first conversion order was for 12 ex-Dutch Mk.6 Hunters converted to T.7 standards as Hunter T. Mk.66D. They were delivered over 1966 and 1967 receiving the registration nos. S.570 to S.581.

A second order was for five ex-R.A.F. F.6 Hunters also converted to T.7 standards as the Hunter T. Mk 66E. They were delivered in 1973 and had registration nos. S1389 to S.1393.

The very last Indian Hunter was phased out in 2001, being replaced by licence-built Su-30MKI's.

Nice Hunter F. Mk. 58 formation of the Swiss air force aerobatic team Patrouille de Suisse, photographed in 1991

A bright red Belgian Hunter F.6 from the aerobatic team 'Diables Rouges' or Red Devils as exhibited in the Brussels Aviation Museum

Singapore- In July 1968 the Republic of Singapore placed a contract for sixteen ex-R.A.F. Mk 6 Hunters. Twelve were built conform FGA 9 standards as Mk. 74. The other four were delivered as FR.Mk.74A fitted with camera's conform FR.10 standards. They were delivered in 1969 and 1970 and entered service at a squadron leaded by former ex-R.A.F. Hunter pilot Chris Strong.

A further order followed for 22 Mk.74B Hunters, converted from F.4 to FGA. 9 standards.

Singapore also ordered four two-seat Hunters as T. Mk. 75. They were refurbished from existing R.A.F. and Dutch KLu T.7's and delivered in 1970. A second order for five additional trainers as T. Mk.75A followed two years later; they were converted from ex-R.A.F. Hunters F.4. Most of the single-seat Hunters from the Singapore air force were later upgraded to carry two Philco Sidewinder missiles under their wings. The Singaporean Hunters were finally phased out in 1992.

South American users

Chile- Chile undertook the acquisition of Hunters from Britain in the 1960s for use in the Chilean Air Force. They purchased 15 Hunter FGA. Mk.71 refurbished from R.A.F., Dutch and Belgian F.6 Hunters. Delivery was completed in 1968. Registrations were J-700 to J-714.

A second order followed in 1973 with 9 more, designated as FGA Mk. 71A. They received the registration numbers J-722 to J730. Last order was in 1974 for another four, converted from F.4 stock with registration numbers J-731, J-733 and J-737.

Chile also ordered six two seat trainers as T. Mk. 72. Also Hawker's famous demonstrator G-APUX went to Chile with this order. Registration nos. were J-718 to J-720 and J-721, J-723 and J-736.

Peru- Peru ordered 16 refurbished ex-R.A.F. Hunters F.4 designated as F. Mk. 52. They were delivered in May 1956 and carried the registration nos. 630 to 645.

Peru also acquired a single two-seater as Hunter T. Mk. 61 with registration number 681.

The Peruvian Hunters were used until they were in 1976 replaced by the Sukhoi Su-22 .

African users

Kenya- Kenya ordered four refurbished ex-R.A.F. Hunters updated to full FGA.9 standards.

They were delivered over 1974-1975 as the FGA. Mk.80 and received the registrations 803 to 806. Kenya also purchased two Hunter trainers as T. Mk. 81. They were converted from ex.-R.A.F. two-seaters and were delivered in 1974. They received the registration numbers 801 and 802. They only served for a

Spencer

Spencer Flack's bright red Hawker Hunter G-HUNT, photographed in September 1980 at Rotterdam airport during an airshow. It is ex-Danish AF Hunter F.4 E-418. Later it was sold to the U.S.A. where it became N611JR but it has not flown since 1982. It was offered for sale for U.S. $ 50,000

Ex Swiss

An ex-Swiss air force Hunter F. Mk. 58 far away from home! It was photographed in the summer of 2000 at the Grand Canyon airport in the U.S.A. Flat nose tyre indicates it was not airworthy

Orbis

A Hunter two-seater at the ILA in 2010, sponsored by Orbis

short period and in 1981 five survivors were sold to Zimbabwe

Somalia- Somalia had nine ex-Abu Dhabi air force Hunters in service in the eighties; eight FGA. Mk.76's and a single trainer T. Mk.77. When the central government collapsed in 1991 all activities by the Somalian air force also came to an end and the Hunters were left to rot away in their revetments.

Zimbabwe (Rhodesia)- During the sixties, the Rhodesian air force a number of Hunter FGA.9's in their fleet that were extensively used to fight against ZANU/ZAOU rebel forces. When Rhodesia became Zimbabwe, a number of these Hunters were 'inherited'. The Air Force of Zimbabwe used its Hunters

to support Laurent Kabila during the Second Congo War in 1998, and were reported to be involved in the fighting in Mozambique. It seems none of the Hunters from the Zimbabwe air force are still operational.

Into battle

Although the Hunter was a typical 'Cold War' product and stationed in Germany near the borders of the Warschau Pact countries it never fired its guns in anger on Soviet aircraft during this period.

During the Suez Campaign ('Operation Musketeer'), in the autumn of 1956, when Egyptian president Nasser occupied the Suez canal zone

the Hunter was operational, but all strike missions were made from aircraft carriers by Westland Wyverns and Hawker Seahawks.

The Hunter was only used for escort missions to provide 'top cover' during these attacks, but it was never confronted with enemy aircraft.

During the early sixties a conflict between the Netherlands and Indonesia over Netherlands New Guinea seemed to escalate from incidents into an actual war. The KLu shipped as an operational trial a small number of Hunters F.4 to Biak. However, not only range was found to be much too short, the Hunters also suffered from the tropical conditions. They were later replaced by the Hunter F.6 fitted with extra fuel tanks under the wings. Even these Hunt-

Aviodrome

At the Aviodrome Theme Park Musem at Lelystad airport Hunter F.4 WV395 is used as a gate guard. It is ex-Danish AF no E-410. It was purchased back from Denmark by Hawker Siddeley and carried for a short period the Class-B marking G-9-438. In 1979 it went to the Netherlands where it was displayed in the earlier Aviodome Museum at Schiphol airport in its present colours

Miss

Hawker Hunter 'Miss Demeanour' photographed in September 2011 at Duxford. 'Miss Demeanour' is registered as G-PSST. It is ex-R.A.F. Hunter F.4 XF974, that later went into service at the Swiss Air Force as J-4104. It is sponsored by a Swiss company called 'Orbis'

ers had a too short range for an effective patrol of the coastal borders and they never saw any action. Long range patrol was taken over by Marineluchvaartdienst Lockheed Neptunes that were appropriately nicknamed as 'Nepfighters'! Eventually Dutch forces withdrew under strong international pressure and Netherlands New Guinea was annexed by the Republic of Indonesia as Irian Jaya.

In 1967, Hunters of the Iraqi Air Force saw action after the Six-Day War between Israel and several neighbouring Arab nations. During the War of Attrition Iraqi Hunters usually operated from bases in Egypt and Syria. While flying a Hunter from Iraqi Airbase H3, Flight Lieutenant Saiful Azam, on exchange from the Pakistan Air Force, shot down two Israeli jets including a Mirage IIIC

Some missions were also flown by the Royal Jordanian Air Force, but most of the Jordanian Hunters were destroyed on the ground on the first day of the Six-Day War.

A Lebanese Hunter shot down an Israeli jet over Kfirmishki in the early 1960s; its pilot was captured by the Lebanese Armed Forces. One Hunter was shot down on the first day of the Six-Day War by the Israeli Air Force. They were used infrequently during the Lebanese Civil War, and eventually fell out of usage and went into storage during the 1980s until being put back in active service in November 2008.

During the Sino-Indian War in 1962, the Hunter's superiority over the Chinese MiG's gave India a strategic advantage; and deterred

the use of bombers from attacking targets within India. The Hunter would also be a major feature in the escalation of the Indo-Pakistani War of 1965; along with the Gnat the Hunter was the primary air defence fighter of India, and regularly engaged in dogfights with Pakistani F-86 Sabres. The aerial war saw both sides conducting thousands of sorties in a single month. Despite the intense fighting, the conflict was effectively a stalemate.

Hunters flown by the IAF extensively operated in the Indo-Pakistani War of 1971; at the start, India had six combat-ready squadrons of Hunters. In the aftermath of the conflict, Pakistan claimed to have shot down 32 Indian Hunters overall. Pakistani infantry and armoured forces attacked the Indian outpost of Longewala in an event now known as the Battle of Longewala. Six IAF Hunters stationed at Jaisalmer Air Force Base were able to halt the Pakistani advance at Longewala by conducting non-stop bombing raids. They attacked Pakistani tanks, armoured personnel carriers and gun positions; and created a sense of chaos on the battlefield, resulting in the Pakistani retreat. Hunters were also used for many ground attack missions and raids into Pakistan, most notoriously used in the bombing of the Attock Oil refinery, to limit Pakistani fuel supplies during the war.

Two years after delivery was completed in 1971, the Hunters were used by military officers in the 1973 Chilean coup d'état to overthrow the socialist president of Chile, Salvador Allende, on 11 September 1973. Coup leaders had ordered the Hunters to relocate to Talcahuano on 10 September. The following morning, they engaged in bombing missions against the presidential palace, Allende's house in Santiago, and several radio stations loyal to the government.

The Royal Rhodesian Air Force used its Hunter FGA.9s extensively against ZANU/ZAPU insurgents in the late 1960s and throughout the 1970s, occasionally engaging in cross-border strikes.

During Siad Barre's regime in present day Somalia, Hunters were used, often flown by former Rhodesian Air Force pilots, to conduct bombing missions during the civil war in the late 1980s.

Survivors and veterans

Even today, in 2012, more than fifty Hunters are still flying with a civil registration.
They are not only active in the airshow circuit, but a number of Hunters are still actively used by civil companies under military contracts. Typical examples are the Hunters from U.S. based ATAC and Apache Aviation operating in Southern France. The Apache Hunters are even used by the French Aeronaval under a civil contract! Well known because of their presence at various airshows are the Swiss Orbis-sponsored Hunters and the two Hunters kept airworthy by the Dutch Hawker Hunter Foundation.
In various aviation museums all over the world the Hunter is well represented and most museums have at least one Hunter in their collection. Typical examples are the Hunter F.6 in Red Devils colours shown the Belgian Aviation Museum at Brussels; the T.7 and F.4 shown in the Militaire Luchtvaart Museum (MLM) at Soesterberg and the ex-Danish air force Hunter F.51 from the Aviodrome at Lelystad airport.

Group Editor	Graphic design
Edwin Hoogschagen	Srecko Bradic
Editors	**Publisher**
Nico Braas	Lanasta
Srecko Bradic	

First print, June 2012
ISBN 978-90-8616-162-1
NUGI 465

Contact Warplane/Violaero:
Slenerbrink 206, 7812 HJ Emmen
The Netherlands
Tel. 0031 (0)591 618 747
info@lanasta.eu

Violaero

References

-Paul Bradley, *The Hawker Hunter - A comprehensive guide*, SAM publications UK (2009)
-J.L. Horsthuis, *Vliegen in Nederland-Hawker Hunter*, Flash Aviation The Netherlands (1990)
-James Goulding, *Interceptor*, Ian Allan UK (1986)
-David J. Griffin, *Hawker Hunter 1951 to 2007*. Lulu Enterprises USA (2006)
-Derek N. James, *Hawker, an aircraft album*, Ian Allan UK (1972)
-Francis K. Mason, *Hawker since 1920 3rd Edition*, Putnam UK (1991)
-Francis K. Mason, *The Hawker Hunter F.6*, Profile Publications no.5, UK (1965)
-Nigel Walpole, *Best of the breed - The Hunter in fighter reconnaissance*, Pen & Sword Aviation UK (2006)

We want to thank the following persons for their contributions to this book:
Mick Gladwin, Daniel Clamot, Jack Cook, Mike Dowsing, David Griffin, Don Hewins, Ian McFarlane, George Trussel and Willem Vredeling

Fokker production

Hawker Hunter F.4 production on 8 March 1956 at Fokker aircraft works at Schiphol